Lecture Notes in Mathematics   1600

Editors:
A. Dold, Heidelberg
F. Takens, Groningen

T0222340

Springer
Berlin
Heidelberg
New York
Barcelona
Budapest
Hong Kong
London
Milan
Paris
Tokyo

Władysław Narkiewicz

Polynomial Mappings

Springer

Author

Władysław Narkiewicz
Institute of Mathematics
Wrocław University
Plac Grunwaldzki 2/4
PL-50-384-Wroclaw, Poland
E-mail: narkiew@math.uni.wroc.pl

Mathematics Subject Classification (1991): 11C08, 11R09, 11T06, 12E05, 13B25, 13F20, 14E05

ISBN 3-540-59435-3 Springer-Verlag Berlin Heidelberg New York

CIP-Data applied for

© Springer-Verlag Berlin Heidelberg 1995
Printed in Germany

Typesetting: Camera-ready output by the author
SPIN: 10130271   46/3142-543210 - Printed on acid-free paper

Preface

1. Our aim is to give a survey of results dealing with certain algebraic and arithmetic questions concerning polynomial mappings in one or several variables. The first part will be devoted to algebraic properties of the ring $Int(R)$ of polynomials which map a given ring R into itself. In the case $R = \mathbf{Z}$ the first result goes back to G.Pólya who in 1915 determined the structure of $Int(\mathbf{Z})$ and later considered the case when R is the ring of integers in an algebraic number field. The rings $Int(R)$ have many remarkable algebraic properties and are a source of examples and counter-examples in commutative algebra. E.g. the ring $Int(\mathbf{Z})$ is not Noetherian and not a Bezout ring but it is a Prüfer domain and a Skolem ring. We shall present classical results in this topic due to G.Pólya, A.Ostrowski and T.Skolem as well as modern development.

2. In the second part we shall deal with *fully invariant sets* for polynomial mappings Φ in one or several variables, i.e. sets X satisfying $\Phi(X) = X$. In the case of complex polynomials this notion is closely related to Julia sets and the modern theory of fractals, however we shall concentrate on much more modest questions and consider polynomial maps in fields which are rather far from being algebraically closed. Our starting point will be the observation that if f is a polynomial with rational coefficients and X is a subset of the rationals satisfying $f(X) = X$, then either X is finite or f is linear. It turns out that the same assertion holds for certain other fields in place of the rationals and also for a certain class of polynomial mappings in several variables. We shall survey the development of these question and finally we shall deal with cyclic points of a polynomial mapping, i.e. with fixpoints of its iterates. Here we shall give the classical result of I.N.Baker concerning cyclic points of complex polynomials and then consider that question in rings of integers in an algebraic number field.

There are several open problems concerning questions touched upon in these lectures and we present twenty one of them.

3. This text is based on a course given by the author at the Karl-Franzens University in Graz in 1991. I am very grateful to professor Franz Halter-Koch for organizing my stay in Graz as well for several very fruitful discussions. My thanks go also to colleagues and friends who had a look at the manuscript and in particular to the anonymous referee who pointed out some inaccuracies.

The work on these lecture notes has been supported by the KBN grant 2 1037 91 01. The typesetting has been done by the author using $\mathcal{AMS}$-TEX.

Notations

We shall denote the rational number field by $\mathbf{Q}$, the field of reals by $\mathbf{R}$, the complex number field by $\mathbf{C}$ and the field of p-adic numbers by $\mathbf{Q}_p$. The ring of rational integers will be denoted by $\mathbf{Z}$, the set of nonnegative rational integers by $\mathbf{N}$, the ring of integers of $\mathbf{Q}_p$ by $\mathbf{Z}_p$, the finite field of q elements by $\mathbf{F}_q$ and the ring of integers in an algebraic number field K by $\mathbf{Z}_K$.

By $a \mid b$ we shall denote the divisibility in various rings and in case of the ring of rational integers we shall write $q \parallel a$ in the case when q is the maximal power of a prime which divides a. The same notation will be used for divisibility of ideals in Dedekind domains. The symbol $\square$ will mark the end of a proof.

CONTENTS

PART A

Rings of integral-valued polynomials

I. Polynomial functions

1. Let R be an arbitrary commutative ring with unit. Every element f of $R[X]$, the ring of all polynomials in one variable with coefficients in R, defines a map $T_f : R \longrightarrow R$. The set of all maps T_f obtained in this way forms a ring, the *ring of polynomial functions on* R, which we shall denote by $P(R)$. Let I_R denote the set of all polynomials $f \in R[X]$ satisfying $f(r) = 0$ for all $r \in R$, and let $F(R)$ denote the set of all maps $R \longrightarrow R$. The following lemma collects a few easy facts concerning $P(R)$ and I_R:

LEMMA 1.1. (i) *The set I_R is an ideal in $R[X]$ and we have*

$$P(R) \simeq R[X]/I_R,$$

(ii) *If R is a domain then the equality $I_R = \{0\}$ holds if and only if R is infinite,*

(iii) *If $R = \mathbf{F}_q$ then I_R is generated by the polynomial $X^q - X$ and*

$$P(R) \simeq R[X]/(X^q - X)R[X].$$

PROOF: The assertion (i) is evident. If R is infinite then clearly only the zero polynomial vanishes identically on R. If R is a finite domain then it is a field, say $R = \mathbf{F}_q$, and the polynomial $X^q - X$ vanishes identically. This proves (ii).

The last assertion follows from the remark that if a polynomial vanishes at all elements of the field $\mathbf{F}_q$ then it must be divisible by

$$\prod_{a \in \mathbf{F}_q} (X - a) = X^q - X. \qquad \square$$

2. The ring $P(R)$ can be described in terms of certain ideals of R:

THEOREM 1.2. (J.WIESENBAUER [82]) *If R is a commutative ring with unit element and for $j = 0, 1, \ldots$ we define I_j to be the set of all $a \in R$ such that there exist $c_0, c_1, \ldots, c_{j-1}$ in R with*

$$ax^j + \sum_{i=0}^{j-1} c_i x^i = 0 \quad \text{for all } x \in R$$

then the I_j's form an ascending chain of ideals in R and if for $j = 0, 1, \ldots$ we fix a set A_j of representatives of R/I_j containing 0, then every $f \in P(R)$ can be uniquely written in the form

$$f(x) = \sum_{j=0}^{N} a_j x^j$$

with a suitable N, $a_j \in A_j$ and $a_N \neq 0$ in case $f \neq 0$.

PROOF: Clearly the I_j's form an ascending chain of ideals. Assume that our assertion fails for some non-zero $f \in P(R)$. Consider all possible polynomial representations:

$$\mathcal{R}: \qquad f(x) = \sum_{j=0}^{m} d_j x^j \quad (x \in R, \ d_j \in R, \ d_m \neq 0)$$

and denote by $i(\mathcal{R})$ the maximal index j with $d_j \notin A_j$. Choose now a representation $\mathcal{R}_0$ with $i = i(\mathcal{R}_0)$ minimal and write

$$(1.1) \qquad f(x) = \sum_{j=0}^{i} d_j x^j + \sum_{j=1+i}^{m} a_j x^j,$$

with $d_i \notin A_i$ and $a_j \in A_j$ for $j = 1+i, 2+i, \ldots, m$. If $a_i \in A_i$ satisfies $a_i - d_i \in I_i$ then with suitable $b_0, b_1, \ldots, b_{i-1} \in R$ we have

$$(a_i - d_i)x^i + \sum_{j=0}^{i-1} b_j x^j = 0 \qquad \text{for all } x \in R,$$

hence in (1.1) we may replace the term $d_i x^i$ by

$$a_i x^i + \sum_{j=0}^{i-1} b_j x^j,$$

contradicting the choice of $\mathcal{R}_0$. $\square$

3. If p is a rational prime then every function $\mathbf{Z}/p\mathbf{Z} \longrightarrow \mathbf{Z}/p\mathbf{Z}$ can be represented by a polynomial. The next theorem describes commutative rings with unit having this property.

THEOREM 1.3. (L.RÉDEI, T.SZELE [47], part I) *Let R be a commutative ring with a unit element. Every function $f : R \longrightarrow R$ can be represented by a* *polynomial from $R[X]$, i.e. $F(R) = P(R)$ holds if and only if R is a finite field.*

PROOF: The sufficiency of the stated condition follows immediately from the interpolation formula of Lagrange, so we concentrate on its necessity.

If R is infinite and its cardinality equals α, then the cardinality of $R[X]$ also equals α but the cardinality of all maps $R \longrightarrow R$ equals $\alpha^\alpha > \alpha$, hence not every such map can be represented by a polynomial.

Let thus $R = \{a_1 = 0, a_2, \ldots, a_n\}$ be a finite unitary commutative ring of n elements. If it is not a field, then it has a zero-divisor c, since every finite domain is necessarily a field. Put

$$g(X) = \prod_{i=1}^{n}(X - a_i),$$

and observe that for all $a \in R$ one has $g(a) = 0$. This shows that if a map $R \longrightarrow R$ can be represented by a polynomial F, then F can be chosen to have its degree $\leq n-1$, since F and F mod g represent the same function on R. The number of all maps $R \longrightarrow R$ and the number of all polynomials of degree $\leq n-1$ both equal n^n and hence it is sufficient to find a non-zero polynomial of degree $\leq n-1$ vanishing on R. The polynomial

$$f(X) = c\prod_{i=2}^{n}(X - a_i)$$

can serve as an example since it evidently vanishes at non-zero arguments and moreover we have $f(0) = (-1)^{n-1}ca_2a_3\cdots a_n$, but as c is a zero-divisor there is an element $a_i \neq 0$ with $ca_i = 0$ and thus $f(0) = 0$. $\square$

(This argument can be modified to cover also rings which do not have a unit element. Cf. L.RÉDEI, T.SZELE [47], part I, p.301).

4. Consider the following example:

Let $R = \mathbf{Z}/4\mathbf{Z}$ be the ring of residue classes mod 4 and put

$$f(x) = \begin{cases} 0 & \text{if } x = 0, 1 \\ 1 & \text{if } x = 2, 3. \end{cases}$$

The function f cannot be represented by a polynomial over R, since otherwise we would have

$$1 \equiv f(3) \equiv f(1) \equiv 0 \pmod 2.$$

However the polynomial

$$g(X) = \left(\frac{X(X-1)}{2}\right)^2$$

attains integral values at integers and it induces on R the function f.

This situation is a special case of the following construction:

Let R and $S_1 \subset S_2$ be commutative rings and let $F : S_1 \longrightarrow R$ be a surjective homomorphism. If a polynomial $f \in S_2[X]$ satisfies

(i) $f(S_1) \subset S_1$,

and

(ii) *If $s, t \in S_1$ and $F(s) = F(t)$ then $F(f(s)) = F(f(t))$,*

then f induces a map $\hat{f} : R \longrightarrow R$ defined by

$$\hat{f}(r) = F(f(s)),$$

where s is any element of S_1 with $F(s) = r$.

Following L.RÉDEI, T.SZELE [47] we shall say that S_2 is a *representation ring* for R, provided there exists $S_1 \subset S_2$ such that every map $g : R \longrightarrow R$ equals $\hat{f}$ for a suitable $f \in S_2[X]$ satisfying (i) and (ii). We shall also say that the pair $< S_1, S_2 >$ is a *representation pair* for R.

THEOREM 1.4. (T.SKOLEM [40]) *If $q = p^k$ is a prime power then $< \mathbf{Z}, \mathbf{Q} >$ is a representation pair for $\mathbf{Z}/q\mathbf{Z}$.*

PROOF: In case $k = 1$ the assertion follows from Theorem 1.3. Assume thus $k \geq 2$. The main step of the proof is embodied in the following lemma:

LEMMA 1.5. *If $q = p^k$ with prime p then there exists a polynomial $\Phi(X) \in \mathbf{Q}[X]$ which is integral-valued at the integers and satisfies*

$$\Phi(x) \equiv \begin{cases} 1 \pmod{q} & \text{if } q \text{ divides } x, \\ 0 \pmod{q} & \text{otherwise.} \end{cases}$$

PROOF: Let $r_1, r_2, \ldots, r_t$ be a complete reduced system of residues mod q and put

$$\Psi(X) = \prod_{i=1}^{t} (X - r_i) \left(\binom{X}{p} - r_i \right) \left(\binom{X}{p^2} - r_i \right) \cdots \left(\binom{X}{p^{k-1}} - r_i \right),$$

$$\Phi(X) = \Psi(X)^2.$$

If $x \in \mathbf{Z}$ is divisible by q, then all numbers

$$\binom{x}{p}, \ldots, \binom{x}{p^{k-1}}$$

are divisible by p. Now observe that if r runs over all residues mod q not divisible by p and a is divisible by p, then $a - r$ runs over all residues mod q not divisible by p and this gives

$$\Psi(x) \equiv (r_1 \cdots r_t)^k \equiv \pm 1 \pmod{q},$$

and

$$\Phi(x) \equiv 1 \pmod{q}.$$

If however q does not divide $x \in \mathbf{Z}$, then we may write $x = p^m y$ with $0 \leq m < k$ and y not divisible by p. Since, as is easily checked, $\binom{x}{p^m}$ is not divisible by p, we have with a suitable i

$$\binom{x}{p^m} \equiv r_i \pmod{q},$$

hence $\Phi(x) \equiv \Psi^2(x) \equiv 0 \pmod{q}$. $\square$

To conclude the proof of the theorem observe that if the map $f : \mathbf{Z}/q\mathbf{Z} \longrightarrow \mathbf{Z}/q\mathbf{Z}$ is arbitrary and a_i is a representative of the residue class $f(i)$ mod q, then the polynomial

$$F(X) = \sum_{i=0}^{q-1} a_i \Phi(X - i)$$

represents f. $\square$

(L.RÉDEI,T.SZELE [47] showed also that the ring of all rational numbers whose denominators are powers of a prime p can serve as a representation ring for $\mathbf{Z}/q\mathbf{Z}$ where q is a power of p. They proved moreover that every ring whose additive group is a cyclic p-group has $\mathbf{Q}$ for its representation ring).

It should be noted that the analogue of Theorem 1.4 fails for composite integers which are not prime-powers. Indeed, assume that one can find a polynomial $f \in S(\mathbf{Z})$ such that

$$f(x) \equiv \begin{cases} 1 \pmod{6} & \text{if 6 divides } x, \\ 0 \pmod{6} & \text{otherwise.} \end{cases}$$

If we write

$$f(X) = \frac{g(X)}{q}$$

with $g \in \mathbf{Z}[X]$ and $q \in \mathbf{Z}$ then

$$g(x) \equiv \begin{cases} q \pmod{6q} & \text{if 6 divides } x, \\ 0 \pmod{6q} & \text{otherwise.} \end{cases}$$

Observe that $6 \mid q$ because if $p = 2$ or $p = 3$ then

$$0 \equiv g(p) \equiv g(6) \equiv q \pmod{p}$$

and $p \mid q$ follows. Write now $q = 2^\alpha M$ with $\alpha \geq 1$ and odd $M \in \mathbf{Z}$ divisible by 3 and choose $x \in \mathbf{Z}$ satisfying

$$x \equiv 0 \pmod{2^{1+\alpha}}, \quad x \equiv 1 \pmod{M}.$$

Then $f(x) \equiv 0 \pmod{6}$ and $f(0) \equiv 1 \pmod{6}$, however $x \equiv 0 \pmod{2^{1+\alpha}}$ implies $g(x) \equiv g(0) \pmod{2^{1+\alpha}}$, hence with a suitable $A \in \mathbf{Z}$ we may write $g(x) - g(0) = A2^{1+\alpha}$ and finally the number

$$f(x) - f(0) = \frac{2A}{M}$$

turns out to be even, contradicting $f(x) - f(0) \equiv 5 \pmod{6}$.

5. We shall see later (see the Corollary to Theorem 1.7) that for composite m, which are not prime powers, the ring $\mathbf{Z}/m\mathbf{Z}$ does not have any representation pair. The problem of determination of all commutative rings which have a representation ring seems to be open (**PROBLEM I**). We shall present now a necessary condition given by L.RÉDEI, T.SZELE [47], but first we have to recall certain elementary properties of difference operators:

If R is an arbitrary commutative ring and $f : \mathbf{Z} \longrightarrow R$ an arbitrary map, then we put

$$\Delta^1 f(x) = f(x+1) - f(x),$$

and

$$\Delta^{n+1} f(x) = \Delta^n f(x+1) - \Delta^n f(x) \quad \text{for } n = 0, 1, \dots .$$

LEMMA 1.6. (i) *For any* $f : \mathbf{Z} \longrightarrow R$, *for* $n = 1, 2, \dots$ *and for all* $x \in \mathbf{Z}$ *one has*

$$\Delta^n f(x) = \sum_{i=0}^{n} (-1)^{n-i} \binom{n}{i} f(x+i),$$

(ii) *If* $f \in R[X]$, $r \in R$ *and we put for* $x \in \mathbf{Z}$

$$g_r(x) = f(xr),$$

then for a suitable positive integer N *we have*

$$\Delta^N g_r(x) = 0$$

for all $x \in \mathbf{Z}$.

PROOF: The assertion (i) is obtained by a simple recurrence argument and to prove (ii) it suffices to observe that $g_r(x)$ is a polynomial in x. □

COROLLARY. *Let* R *be a ring having a representation ring and let* $< S_1, S_2 >$ *be a representation pair for* R. *For every map* $F : R \longrightarrow R$ *and every non-zero* $r \in R$ *there exists a positive integer* N *such that the* N-*th iterate* δ_r^N *of the operator* δ_r, *defined by*

$$\delta_r F(x) = F(x+r) - F(x)$$

vanishes identically.

PROOF: Let $\varphi : S_1 \longrightarrow R$ be a surjective homomorphism, realizing the representation of R by the pair $< S_1, S_2 >$ and let $r = \varphi(s)$ for some $s \in S_1$. It suffices now to apply part (ii) of the lemma to the polynomial inducing F. □

THEOREM 1.7. (L.RÉDEI, T.SZELE [47], part II, Satz 5) *Let* R *be a commutative ring with unit element* e *and assume that* R *has a representation ring. Then there exists a prime power* q *such that* $qe = 0$.

PROOF: Let $< S_1, S_2 >$ be a representation pair for R.

First assume that R contains an element s of infinite additive order, i.e. all elements $s, 2s, 3s, \dots$ are distinct and non-zero, and let $f : R \longrightarrow R$ satisfy $f(s) = s$ and $f(ks) = 0$ for $k \in \mathbf{Z}$. It suffices now to apply the Corollary to Lemma 1.6, since it is obvious that none of the iterated differences of the sequence $s, 0, 0, \ldots$ can vanish.

We may thus assume that all elements of R have a finite additive order. Assume also that there is a non-zero element $s \in R$ whose order m is not a prime power. Let p be a prime divisor of m and define k by $q = p^k \parallel m$. Consider any map $f : R \longrightarrow R$ satisfying $f(is) = e$ for $i \equiv p^k \pmod{m}$ and $f(is) = 0$ for all other $i \in \mathbf{Z}$. By the Corollary to Lemma 1.6 the N-th differences $\delta_r^n f$ vanish for all sufficiently large N and hence we may find such an N which is a power of p, say $N = p^u$. We may assume that u exceeds k and moreover the congruence

$$p^{u-k} \equiv 1 \pmod{\frac{m}{p^k}}$$

holds. (Simply choose sufficiently large u satisfying $u \equiv k \pmod{\varphi(m/p^k)}$). The last congruence implies

$$(1.2) \qquad\qquad p^u \equiv p^k \pmod{m}$$

and if we put $\hat{f}(x) = f(rx)$ then with the use of Lemma 1.6 (i) we get

$$\hat{f}(p^u) - \binom{p^k}{1} \hat{f}(p^u - 1) + \cdots + (-1)^{p^k} \hat{f}(0) = 0$$

and the congruence

$$\hat{f}(p^u) \equiv \hat{f}(0) \pmod{pR}$$

follows. (Note that our assumption about m implies that pR is not the zero ideal). Finally note that since the function $\hat{f}$ is periodic of period m, the congruence (1.2) leads to

$$0 = \hat{f}(0) \equiv \hat{f}(p^u) \equiv \hat{f}(p^k) \pmod{pR},$$

a contradiction.

It follows that the additive order of every non-zero element of R must be a prime-power, and this applies in particular to the unit element. $\square$

COROLLARY. *The ring $\mathbf{Z}/m\mathbf{Z}$ has a representation ring if and only if m is a prime power.*

PROOF: The necessity follows from the last theorem and the sufficiency is contained in Theorem 1.4. $\square$

6. We conclude this section with two results dealing with $P(R)$ in the case $R = \mathbf{Z}/q\mathbf{Z}$ and start with a theorem of L.CARLITZ [64]:

THEOREM 1.8. *Let $q = p^n$ be a prime power, let $f : \mathbf{Z}/q\mathbf{Z} \longrightarrow \mathbf{Z}/q\mathbf{Z}$ be a given map, let $A_q = \{0, 1, 2, \ldots, q - 1\}$ and denote by $\hat{f} : A_q \longrightarrow A_q$ the map induced by f. Then $\hat{f}$ is a restriction to A_q of a polynomial $F \in \mathbf{Z}[X]$ if and only if $\Delta^r \hat{f}(0)$ is divisible by $(q, r!)$ for $r = 0, 1, \ldots, q - 1$.*

PROOF: *Necessity.* In view of Lemma 1.6 (i) it suffices to establish the following lemma:

LEMMA 1.9. *If $F \in \mathbf{Z}[X]$ then the numbers*

$$\sum_{i=0}^{r}(-1)^{r-i}\binom{r}{i}F(i)$$

are for all $r \geq 0$ divisible by $r!$.

PROOF: Observe first that if we define for $j = 0, 1, \ldots$ the polynomials F_j by

$$F_j(X) = X(X-1)\cdots(X-j+1),$$

then every polynomial of $\mathbf{Z}[X]$ can be uniquely written as a linear combination of the F_j's with rational integral coeffficients. A simple inductive argument shows that for $r = 1, 2, \ldots$ one has

$$\Delta^r F_j(X) = j(j-1)\cdots(j-r+1)F_{j-r}(X),$$

and since the product of r consecutive integers is divisible by $r!$ the assertion follows for the polynomials F_j and by linearity the lemma results. $\square$

Sufficiency. We need a simple lemma:

LEMMA 1.10. *If h is any function defined on the set A_q then for for all $a \in A_q$ one has*

$$h(a) = \sum_{j=0}^{q} \Delta^j h(0)\binom{a}{j}.$$

PROOF: Using Lemma 1.6 (i) and the equality

$$\binom{a}{j}\binom{j}{i} = \binom{a}{i}\binom{a-i}{j-i}$$

we get

$$\sum_{j=0}^{q}\Delta^j h(0)\binom{a}{j} = \sum_{i=0}^{q}h(i)\sum_{j=i}^{q}(-1)^{j-i}\binom{j}{i}\binom{a}{j}$$

$$= \sum_{i=0}^{q}\binom{a}{i}h(i)\sum_{t=0}^{q-i}(-1)^t\binom{a-i}{t}.$$

Since for $t > a - i$ we have $\binom{a-i}{t} = 0$, the last expression equals

$$\sum_{i=0}^{q}\binom{a}{i}h(i)\sum_{t=0}^{a-i}(-1)^t\binom{a-i}{t} = \sum_{i=0}^{q}\binom{a}{i}h(i)(1-1)^{a-i} = h(a). \quad \square$$

Observe now that the denominator of the fraction $\Delta^j \hat{f}(0)/j!$ in its reduced form is for $j \leq q$ not divisible by p and so we may find $0 \leq \xi_j < q$ satisfying

$$\Delta^j \hat{f}(0) \equiv \xi_j j! \pmod{q}.$$

Applying the last lemma to $h = \hat{f}$ we obtain that the polynomial

$$F(X) = \sum_{j=0}^{q} \xi_j X(X-1)\cdots(X-j+1)$$

realizes $\hat{f}$. □

It has been shown by F.DUEBALL [49] that if p is a prime and $n > 1$ then every polynomial in $\mathbf{Z}/(p^n\mathbf{Z})[X]$ is uniquely determined by its values at $x = 0, 1, \ldots, tp - 1$, where t is defined as follows: if $p^{c_j} \parallel p^j j!$ for $j = 0, 1, \ldots$, then c is the smallest index satisfying $c_t \geq n$. This is closely related to the polynomial interpolation problem. A necessary and sufficient condition for its solvability in an arbitrary commutative ring has been given by R.SPIRA [68].

A characterization of functions $f : (\mathbf{Z}/p^n\mathbf{Z})^k \longrightarrow \mathbf{Z}/p^n\mathbf{Z}$ which can be represented by k-ary polynomials has been given by I.G.ROSENBERG [75].

7. The number of of elements of $P(\mathbf{Z}/q\mathbf{Z})$ for any integer q has been found by A.J.KEMPNER [21]. We give a proof due to J.WIESENBAUER [82]. (Another proofs had been given by G.KELLER, F.R.OLSON [68] and G.MULLEN, H.STEVENS [84]. Cf. also J.V.BRAWLEY, G.L.MULLEN [92], who considered the more general case of polynomial functions in a *Galois ring* $\mathbf{Z}[X]/I$ with $I = p\mathbf{Z}[X] + f\mathbf{Z}[X]$, where p is a prime and $f \in \mathbf{Z}[X]$ is a polynomial irreducible mod p. For the theory of Galois rings see [MD]. The number of elements in $P(R)$ in the case when R is a finite commutative local principal ideal ring has been determined by A.A.NEČAEV [80]).

THEOREM 1.11. *Let* $n \geq 1$ *be an integer and denote by* $M(n)$ *the cardinality of* $P(\mathbf{Z}/n\mathbf{Z})$. *Then*

$$M(n) = \prod_{i=0}^{N} \frac{n}{(n, N!)} \quad ,$$

where $N = N_n$ *denotes the largest integer such that* n *does not divide* $N!$. *Moreover* $M(n)$ *is a multiplicative function, i.e.* $(n_1, n_2) = 1$ *implies* $M(n_1 n_2) = M(n_1)M(n_2)$.

In particular, if $n = p^k$ *is a prime power, then* $M(n) = p^{k(N+1)-s}$ *where* s *is the exponent of the prime* p *in the canonical factorization of* $\prod_{j=2}^{N} j!$.

PROOF: It has been noted in the proof of Lemma 1.9 that every polynomial f of degree not exceeding r over $\mathbf{Z}$ can be written uniquely in the form

$$f(X) = \sum_{j=0}^{r} a_j F_j(X),$$

with $F_j(X) = X(X-1)\cdots(X-j+1)$. Restricting the coefficients by $0 \leq a_j < n$ we get a general form of a polynomial over $\mathbf{Z}/n\mathbf{Z}$. Observe now that f has all its values divisible by n if and only if for $i = 0, 1, \ldots, r$ one has

$$a_i \equiv \frac{n}{(n, i!)} \pmod{n}.$$

In fact, if this condition is satisfied, then we get

$$a_i f(X) = a_i i! \binom{X}{i} \equiv n \frac{i!}{(n, i!)} \binom{X}{i} \equiv 0 \pmod{n},$$

and if f vanishes identically mod n, then evidently $a_0 = f(0)$ is divisible by n and if for $k = 0, 1, \ldots, i-1$ we have $a_k \equiv n/(n, k!) \pmod{n}$, then

$$0 \equiv f(i) = \sum_{j=0}^{r} a_j f_j(i) \equiv a_i j! \pmod{n},$$

implying our assertion. It follows that for every j the ideal I_j occuring in theorem 1.2 is generated by $n/(n, j!)$ and thus the first assertion follows from that theorem. Multiplicativity of $M(n)$ is an immediate consequence of the Chinese Remainder Theorem and the last assertion is just a special case of the first. $\square$

Exercises

1. (I.NIVEN, LEROY J.WARREN [57]) Let m be a positive integer and $R = \mathbf{Z}/m\mathbf{Z}$. Prove that I_R is a finitely generated ideal in $R[X]$ which is principal if and only if m is a prime.

2. Let R be a domain and let A be a finite subset of R. Prove that every map $A \longrightarrow R$ can be realized by a polynomial in $R[X]$ if and only if every non-zero difference of elements of A is invertible.

3. Prove the analogue of Theorem 1.3 for functions of several variables.

4. (G.MULLEN, H.STEVENS [84]) Prove the analogue of Theorem 1.11 for polynomials in several variables.

5. Let $f \in \mathbf{Z}[X]$ and let N be a positive integer. One says that f is a *permutation polynomial* mod N, if it induces a permutation of $\mathbf{Z}/N\mathbf{Z}$.

(i) Show that if p is a prime then $f \in \mathbf{Z}[X]$ is a permutation polynomial mod p^2 if and only if it is a permutation polynomial mod p and for all $x \in \mathbf{Z}$ one has
$$f'(x) \not\equiv 0 \pmod{p}.$$

(ii) Show that if p is a prime and f is a permutation polynomial mod p^2, then it also a permutation polynomial mod p^n for $n = 3, 4, \ldots$.

(iii) Prove that f is a permutation polynomial mod N if and only if it is a permutation polynomial mod q for all prime powers q dividing N.

6. (G.MULLEN, H.STEVENS [84]) Let p be a prime and $n \geq 2$. Prove that the number of polynomial functions which permute the elements of $\mathbf{Z}/p^n\mathbf{Z}$ equals
$$p!(p-1)^p p^D,$$

where

$$D = (N + 1)(n - c(N)) + \delta(N) - 2p,$$

(with N being the largest integer with $c(N) < n$, where $c(N)$ is the exponent of p in the factorization of $n!$) and

$$\delta(n) = \frac{1}{2} \sum_{r=1}^{\infty} p^r \left[\frac{n}{p^r}\right] \left(\left[\frac{n}{p^r}\right] + 1\right).$$

7. (L.CARLITZ [63]) Prove that if a polynomial $f \in \mathbf{F}_p[X]$ induces a permutation in all fields $\mathbf{F}_{p^k}$ ($k = 1, 2, \ldots$) then with suitable $a, b \in \mathbf{F}_p$, $a \neq 0$ and $r \geq 1$ one has

$$f(X) = aX^{p^r} + b.$$

II. The ring Int(R)

1. Let R be an integral domain and K its quotient field. For any subset A of R we shall consider the set $Int(A, R)$ of all polynomials over K mapping A into R. Obviously $Int(A, R)$ is always a ring satisfying $R[X] \subset Int(A, R) \subset K[X]$. Of special importance will be the ring $Int(R, R)$ which we shall for shortness denote by $Int(R)$.

We consider first the classical case $R = \mathbf{Z}$ and prove an elementary result due to G.Pólya [15],[19] and T.Nagell [18] (cf. also A.J.Kempner [21], L.Rédei, T.Szele [47]):

Theorem 2.1. *A polynomial $g \in \mathbf{Q}[X]$ of degree n maps the ring of rational integers $\mathbf{Z}$ into $\mathbf{Z}$ (i.e. is integer-valued) if and only if with suitable $a_0, \dots, a_n \in \mathbf{Z}$ we have*

$$g(X) = \sum_{i=0}^{n} a_i h_i(X),$$

where $h_0(X) = 1$ and

$$(2.1) \qquad h_i(X) = \frac{X(X-1)\cdots(X-i+1)}{i!}$$

for $i = 1, 2, \dots$. This representation is unique.

Proof: We need a simple lemma:

Lemma 2.2. *Let R be an integral domain, K its quotient field and let f be a polynomial over K of degree m lying in $Int(R)$. If A denotes its leading coefficient, then $m!A \in R$.*

Proof: Observe that if we put

$$g_1(X) = g(X+1) - g(X) = mAX^{m-1} + \dots$$
$$g_2(X) = g_1(X+1) - g_1(X) = m(m-1)AX^{m-2} + \dots$$

$$\cdots \quad \cdots \quad \cdots$$

$$g_m(X) = g_{m-1}(X+1) - g_{m-1}(X) = m!A,$$

then each polynomial $g_i(X)$ maps R into R and so in particular $m!A$ must lie in R. $\square$

(A generalization of this lemma to the case of polynomials in several variables has been given in H.Gunji, D.L.McQuillan [69]).

The sufficiency of the condition given in the theorem follows from the observation that for integral $x > i$ the value $h_i(x)$ coincides with the binomial coefficient $\binom{x}{i}$ and thus is integral, for $x = 0, 1, \ldots, i - 1$ we have $h_i(x) = 0$ and for negative x we have $h_i(x) = (-1)^i h_i(-x + i - 1) \in \mathbf{Z}$.

To prove its necessity observe first that the assertion is obvious for constant and linear polynomials. Now assume that it holds for all polynomials of degree not exceeding $m - 1$ and let the polynomial

$$g(X) = c_m X^m + \cdots + c_0$$

(with rational c_i and $c_m \neq 0$) be integer-valued. The lemma implies that $m!c_m$ is an integer, i.e. we have $c_m = b_m/m!$ with a suitable non-zero integer b_m. The polynomial

$$g(X) - b_m h_m(X)$$

is integral-valued and since its degree does not exceed $m - 1$ we may apply to it the inductional assumption and our assertion follows.

The uniqueness results from the fact that the degrees of the polynomials h_i are all distinct. $\square$

The argument given above leads to a slightly stronger statement to the effect that if f is a polynomial with complex coefficients and $E = f(\mathbf{Z})$, then f can be written as a linear combination of the h_i's with coefficients lying in the additive group generated by E.

COROLLARY 1. Let q be a given integer. A polynomial $f \in \mathbf{Q}[X]$ maps $\mathbf{Z}$ in $q\mathbf{Z}$ if and only if f can be written in the form given in the theorem with all coefficients a_i divisible by q. $\square$

The following extension of Theorem 2.1 has been obtained by T.NAGELL [19]:

COROLLARY 2. A polynomial $f \in \mathbf{Q}[X_1, \ldots, X_n]$ attains integral values at points with integral coordinates if and only if it can be written as a linear combination with integral coefficients of polynomials of the form

$$h_{i_1}(X_1) \cdots h_{i_n}(X_n) \quad (i_1, i_2, \ldots, i_n = 0, 1, 2, \ldots).$$

PROOF: The sufficiency of this condition follows from Theorem 2.1 and to prove its necessity we apply induction in n. For $n = 1$ the assertion is a part of Theorem 2.1. Assume its truth for polynomials in $n - 1$ variables for some $n \geq 2$ and let $f \in \mathbf{Q}[X_1, \ldots, X_n]$ be integer-valued.

Since $h_{i_1}(X_1) \cdots h_{i_n}(X_n)$ is of degree i_j in X_j we can write f in a unique way in the form

$$f(X_1, \ldots, X_n) = \sum_{i_1, \ldots, i_n} c(i_1, \ldots, i_n) h_{i_1}(X_1) \cdots h_{i_n}(X_n),$$

with rational coefficients $c(i_1, \ldots, i_n)$ and hence with a suitable N we get

$$f(X_1, \ldots, X_n) = \sum_{j=0}^{N} g_j(X_1, \ldots, X_{n-1}) h_j(X_n)$$

where

$$g_j(X_1, \ldots, X_{n-1}) = \sum_{i_1, \ldots, i_{n-1}} c(i_1, \ldots, i_{n-1}, j) h_{i_1}(X_1) \cdots h_{i_{n-1}}(X_{n-1}).$$

For all $a_1, \ldots, a_{n-1} \in \mathbf{Z}$ the polynomial $f(a_1, \ldots, a_{n-1}, X) \in \mathbf{Q}[X]$ is integer-valued, hence Theorem 2.1 gives $g_j(a_1, \ldots, a_{n-1}) \in \mathbf{Z}$ for $j = 0, 1, \ldots, N$ and using the inductional assumption we obtain $c(i_1, \ldots, i_{n-1}, j) \in \mathbf{Z}$ for all $i_1, \ldots, i_{n-1}, j$. $\square$

2. The assumption $g \in \mathbf{Q}[X]$ in Theorem 2.1 can be relaxed. In fact it has been shown by G.Pólya [15], [20] that if f is an entire function f which maps the non-negative integers into $\mathbf{Z}$ and satisfies

$$\lim_{r \to \infty} \frac{r^{1/2} M(r)}{2^r} = 0,$$

(where $M(r)$ denotes the maximal value of $|f(z)|$ on the circle $|z| = r$), then f is necessarily a polynomial. Note that the example $f(z) = 2^z$ shows that one cannot replace here 2^r by c^r for some $c > 2$.

Pólya's conjecture that the factor $r^{1/2}$ in his theorem can be removed has been confirmed by G.H.Hardy [16] and E.Landau [20] gave a simpler proof. (Cf. D.Sato, E.G.Straus [64]).

Pólya dealt in his paper [15] also with entire functions f assuming integral values at all integer points, i.e. mapping $\mathbf{Z}$ into $\mathbf{Z}$. He showed that if $\theta = \dfrac{3 + \sqrt{5}}{2}$ and

$$\lim_{r \to \infty} \frac{r^{3/2} M(r)}{\theta^r} = 0$$

holds then f must be a polynomial. One cannot replace here θ by a larger number, because the function

$$f(z) = \frac{1}{\sqrt{5}} \left(\theta^z - \theta^{-z} \right)$$

maps $\mathbf{Z}$ in $\mathbf{Z}$.

It should be noted that for every positive δ there exist entire transcendental functions f satisfying both $f(\mathbf{Q}) \subset \mathbf{Q}$ and

$$\lim_{r \to \infty} \frac{M(r)}{\delta^r} = 0,$$

as shown by P.Stäckel [95].

If f is an entire function and

$$\alpha = \limsup_{r \to \infty} \frac{\log M(r)}{r}$$

is finite, then f is said to be of type α. Thus entire functions f of type $< \log 2$ with $f(\mathbf{N}) \subset \mathbf{Z}$ and entire functions g of type $< \log \theta$ with $g(\mathbf{Z}) \subset \mathbf{Z}$ are polynomials. Generalizing previous results of G.PÓLYA [20], F.CARLSON [21] and A.SELBERG [41a],[41c], C.PISOT [42] showed that if f is entire of order not exceeding $\alpha_0 = 0.843\ldots$ and $f(\mathbf{Z}) \subset \mathbf{Z}$, then

$$f(z) = \sum_{j=1}^{k} c_j^z P_j(z),$$

where P_j are polynomials and c_j are certain algebraic numbers.

For further generalizations see A.O.GELFOND [29a], A.SELBERG [41b], C. PISOT [46a], [46b], R.C.BUCK [48], R.WALLISSER [69], R.M.ROBINSON [71], F. GRAMAIN, M.MIGNOTTE [83].

For similar results about entire functions mapping the ring of algebraic integers of a given field into itself see S.FUKASAWA [26], [28], A.O.GELFOND [29], M.WALDSCHMIDT [78],L.GRUMAN [80], D.MASSER [80], F.GRAMAIN [80], [81].

An analogue of Pólya's result for entire functions which for a given integer $q \geq 2$ map the set of all positive integral powers of q in $\mathbf{Z}$ has been obtained by A.O. GELFOND [33]. (See also A.O.GELFOND [67], R.WALLISER [85] and F.GRAMAIN [90]).

For analogues in the case of several variables see A.BAKER [67], V.AVA-NISSIAN, R.GAY [75], F.GRAMAIN [78a], J.-P.BÉZIVIN [84] and P.BUNDSCHUH [80].

Surveys of this topic could be find in F.GRAMAIN [78a], [78b], F.GRAMAIN, F.J.SCHNITZER [89].

3. Theorem 2.1 shows that the polynomials h_i form a set of free generators of $Int(\mathbf{Z})$ as a $\mathbf{Z}$-module. The same assertion may hold also for certain rings other than $\mathbf{Z}$ as we now show.

THEOREM 2.3. *If p is a rational prime then a polynomial $f \in \mathbf{Q}_p[X]$ lies in $Int(\mathbf{Z}_p)$ if and only if it can be written in the form*

$$f(X) = \sum_{i=0}^{n} a_i h_i(X),$$

where the polynomials h_i are given by (2.1) and the coefficients a_i lie in $\mathbf{Z}_p$.

PROOF: We need a simple lemma:

LEMMA 2.4. (P.-J.CAHEN, J.-L.CHABERT [71]) *Let R be a Noetherian domain and let K be its field of quotients. Let P be a maximal ideal of R and denote* by R_P the corresponding localization, i.e.

$$R_P = A^{-1}R = \{\frac{a}{b} : a, b \in R, b \in A\},$$

where $A = R \setminus P$. Then

$$Int(R_P) = A^{-1}Int(R),$$

i.e. a polynomial $f \in K[X]$ lies in $Int(R_P)$ if and only if for some $h \in Int(R)$ and $b \in A$ we have

$$f(X) = \frac{h(X)}{b}.$$

PROOF: We start with the inclusion

$$A^{-1}Int(R) \subset Int(R_P).$$

Let $f(X) = \sum_{j=0}^{N} a_j X^j \in R[X]$, $\Delta \in R$ and assume that $g(X) = f(X)/\Delta \in Int(R)$, thus $f(R) \subset \Delta R$. It suffices to show that for any $a/b \in R_P$ ($a, b \in R$, $b \notin P$) one has $f(a/b) \in \Delta R_P$. Since R_P is Noetherian, there exists an integer $m \geq 0$ such that the principal ideal generated by Δ in R_P contains P^m. Let $u \in R$ be any solution of the congruence

$$a \equiv bu \pmod{P^m}.$$

Writing $a = bu + r$ with $r \in P^m$ we get

$$f\left(\frac{a}{b}\right) = \sum_{j=0}^{N} a_j \left(u + \frac{r}{b}\right)^j \in f(u) + rR_P \subset \Delta R + P^m R_P \subset \Delta R_P,$$

since by construction $P^m R_P \subset \Delta R_P$.

To obtain the converse inclusion assume $f \in Int(R_P)$. We can write

$$f(X) = \frac{g(X)}{q}$$

with a certain $g \in R[X]$ and $q \in R$. If $q \notin P$ then our assertion results immediately, hence assume $q \in P$. If I denotes the ideal of R generated by $g(R)$ then $I \subset P$. Let $g_1, g_2, \ldots, g_m$ be a set of generators of I. With a suitable integer t and $x_j, c_{ij} \in R$ we have

$$g_i = \sum_{j=1}^{t} c_{ij} g(x_j),$$

and by our asssumptions we can write $f(x_j) = g(x_j)/q = a_j/b_j$ for $j = 1, 2, \ldots, t$ and $a_j, b_j \in R$, $b_j \notin P$. If now x is an arbitrary element of R, then with suitable $r_1, r_2, \ldots, r_k$ in R we have

$$g(x) = \sum_{i=1}^{k} r_i g_i = \sum_{i=1}^{k} r_i \sum_{j=1}^{t} c_{ij} q a_j/b_j$$

and hence

$$f(x) = \sum_{i=1}^{k} r_i \sum_{j=1}^{t} c_{ij} a_j / b_j.$$

Putting now $b = \prod_{j=1}^{t} b_j$ we get $b \in A$ and $bf(X) \in Int(R)$, as asserted. $\square$

COROLLARY. *If p is a rational prime and we denote by $\mathbf{Z}_{(p)} = \mathbf{Q} \cap \mathbf{Z}_p$ the corresponding localization of $\mathbf{Z}$, then every polynomial in $Int(\mathbf{Z}_{(p)})$ can be written in the form*

$$\frac{F(X)}{q}$$

with $F \in Int(\mathbf{Z})$ and q not divisible by p.

PROOF: The assertion is a special case of the Lemma. $\square$

Since $\mathbf{Z}$ is dense in $\mathbf{Z}_p$ it follows that all polynomials h_i map $\mathbf{Z}_p$ in $\mathbf{Z}_p$. Now assume that the polynomial $f \in \mathbf{Q}_p[X]$ of degree N maps $\mathbf{Z}_p$ in $\mathbf{Z}_p$ and let f_n be a sequence of polynomials with rational coefficients, satisfying $\deg f_n = N$ and tending to f in the p-adic topology. Observe that for sufficiently large n we have $f_n \in Int(\mathbf{Z}_{(p)})$. Indeed, if for infinitely many n one could find $x_n \in \mathbf{Z}_{(p)}$ with $f_n(x_n) \notin \mathbf{Z}_{(p)}$, then by choosing a suitable subsequence we might assume that $x_0 \in \mathbf{Z}_p$ is the limit of the sequence $\{x_n\}$ and then we would get

$$\lim_{n \to \infty} f_n(x_n) = f(x_0) \notin \mathbf{Z}_p$$

since $\mathbf{Z}_p$ is both closed and open, and this gives a contradiction.

From the Corollary to Lemma 2.4 we infer now that for sufficiently large n one has

$$f_n(X) = \frac{g_n(X)}{q_n}$$

with $g_n \in Int(\mathbf{Z})$ and $q_n \in \mathbf{Z} \setminus p\mathbf{Z}$. Using Theorem 2.1 we write now

$$f_n(X) = \frac{\sum_{j=0}^{N} c_j^{(n)} h_j(X)}{q_n}$$

with suitable $c_j^{(n)} \in \mathbf{Z}$ and the assertion follows by letting n tend to infinity. $\square$

Another example has been given by G.GERBOUD [88a]:

Let K be an algebraic number field, $\mathbf{Z}_K$ its ring of integers, let G be the multiplicative group generated by all rational primes which do not split in $K/\mathbf{Q}$ and let $R = G^{-1}\mathbf{Z}_K$ be the ring of fractions of $\mathbf{Z}_K$ with respect to G. Then the R-module $Int(R)$ is generated by the polynomials h_i given by (2.1).

4. A proof of the following description of domains R having the same property can be found in F.HALTER-KOCH, W.NARKIEWICZ [92c]:

If R is a domain with quotient field K of zero characteristics then $Int(R)$ coincides with the R-submodule of $K[X]$ generated by $h_0, h_1, \ldots$ if and only if *for every rational prime p which is not invertible in R every prime ideal P of R containing p is of index p and one has $PR_P = pR_P$.*

In case of Noetherian domains this condition takes a simpler form:

THEOREM 2.5. (F.HALTER-KOCH, W.NARKIEWICZ [92c]) *If R is a Noetherian domain, then the following conditions are equivalent:*

(i) *$Int(R)$ is generated by $h_0, h_1, \ldots,$*

(ii) *For every rational prime p which is not invertible in R the principal ideal generated by p is a product of distinct maximal ideals of index p in R.*

PROOF: Let K be the quotient field of R, denote by $A(R)$ the R-module generated by the polynomials $h_0, h_1, h_2, \ldots$ and observe (as done in G.GERBOUD [88a]) that $Int(R) \subset A(R)$ holds for any domain R of zero characteristics. Indeed, any polynomial $f \in Int(R)$ of degree N can be written in a unique way as

$$f = \sum_{j=0}^{N} a_j h_j,$$

with $a_j \in K$, $a_m \neq 0$. Now obviously $a_0 = f(0) \in R$ and if $a_0, a_1, \ldots, a_r$ lie in R then

$$a_{r+1} = f(r+1) - \sum_{j=0}^{r} a_j h_j(r+1) \in R,$$

and $f \in A(R)$ follows.

Thus (i) is equivalent to

(iii) *For $i = 1, 2, \ldots$ one has $h_i(R) \subset R$.*

Assume first that (iii) holds and let p be a rational prime which is not invertible in R. Since R is Noetherian we can write

$$pR = I_1 \cap I_2 \cap \cdots \cap I_n$$

where I_j are distinct primary ideals. Denote by P_j the prime ideal with respect to which I_j is primary. Our assumptions imply that for all $x \in R$ one has

$$x(x-1)\cdots(x-p+1) \in pR \subset I_j \subset P_j$$

and thus the set $\{0, 1, \ldots, p-1\}$ must contain a complete residue system mod p and $\#R/P_j = p$ follows, implying the maximality of P_j. If for some j we have $I_j \neq P_j$, then $\#R/I_j > p$ and so we may choose $x \in R$ not congruent to $0, 1, \ldots, p-1 \mod I_j$. Since

$$x(x-1)\cdots(x-p+1) \equiv 0 \pmod{I_j}$$

a certain element $x - k$ $(0 \leq k < p)$ must be zero-divisor in R/I_j. As I_j is P_j-primary, $x - k \in P_j$ follows and this implies that for $i \neq k$ we have $x - i \notin P_j$. If

$$A_k = \prod_{i \neq k} (x-i)$$

then $A_k \notin P_j$, thus A_k cannot be a zero-divisor in R/I_j and in view of

$$A_k(x - k) \equiv 0 \pmod{I_j}$$

we get $x - k \in I_j$ thus $x \equiv k \pmod{I_j}$, contrary to our choice of x. Thus we must have $\#R/I_j = p$ and the equality $I_j = P_j$ follows. Since the I_j's are distinct maximal ideals, their intersection is equal to their product and (ii) follows.

Now assume that (ii) is satisfied and let p be a rational prime not invertible in R. Thus

$$pR = M_1 \cdots M_n$$

holds with certain distinct maximal ideals $M_1, M_2, \ldots, M_n$ of index p in R. An easy induction gives now $\#R/M_j^k = p^k$ for $j = 1, 2, \ldots, n$ and $k = 1, 2, \ldots$. We show now that the set $\{0, 1, \ldots, p^k - 1\}$ forms a complete system of representatives of residue classes mod M_j^k. It suffices to show that these numbers are all distinct mod M_j^k. If this fails, then there is an integer $1 \le i < p^k$ lying in M_j^k. Write $i = \alpha p^l$ with $0 \le l < k$ and α not divisible by p. Then

$$\alpha M_1^l \cdots M_n^l \subset M_j^k.$$

As $\alpha \notin M_j$ the ideals $\alpha M_1^l \cdots M_{j-1}^l M_{j+1}^l \cdots M_n^l$ and M_j^k are relatively prime, thus

$$M_j^k + \alpha M_1^l \cdots M_{j-1}^l M_{j+1}^l \cdots M_n^l = R$$

whence

$$M_j^k M_j^l + \alpha (M_1 \cdots M_n)^l = M_j^l$$

and we get

$$M_j^l \subset M_j^{k+l} + M_j^k \subset M_j^k$$

which leads to $l \ge k$, contradiction.

Now let $j \ge 0$, let p be a prime not exceeding j, let r be the highest power of p dividing $j!$ and let $x \in R$. If p is not invertible in R we use the preceding observation to find for $k = 1, 2, \ldots, n$ a rational integer s_k satisfying $x \equiv s_k \pmod{M_k^r}$. If we put

$$B = s_k(s_k - 1) \cdots (s_k - j + 1)$$

then

$$x(x - 1) \cdots (x - j + 1) \equiv B \pmod{M_k^r},$$

but evidently $B \in j!R \subset p^r R \subset M_k^r$ holds for $k = 1, 2, \ldots, n$ thus

$$x(x - 1) \cdots (x - j + 1) \in (M_1 \cdots M_n)^r = p^r R.$$

The last inclusion being evident for p invertible, (iii) follows. $\square$

In case of a Dedekind domain Theorem 2.5 has been also obtained by J. - L.CHABERT, G.GERBOUD [93]. For a generalization see G.GERBOUD [93].

5. If $A \subset \mathbf{Z}$, then obviously $Int(\mathbf{Z}) \subset \mathbf{Q}(A, \mathbf{Z})$. A description of those sets A of integers for which $\mathbf{Q}(A, \mathbf{Z})$ coincides with $Int(\mathbf{Z})$ has been given by

R.GILMER [89], who proved the following result:

THEOREM 2.6. *Let A be a set of integers. The ring $Q(A, \mathbf{Z})$ coincides with $Int(\mathbf{Z})$ if and only if for every prime-power q the set A contains elements from every residue class mod q.*

(Sets with this property are called *arithmetically dense*. See G.RAUZY [67]).

PROOF: Assume first that A contains a complete set of representatives of residue classes mod q for every prime-power q and let f be a polynomial assuming integral values on A. If we write $f(X) = g(X)/N$ with integral N and $g \in \mathbf{Z}[X]$ then clearly $g(A) \subset N\mathbf{Z}$. If now x is an integer and q is any prime-power divisor of N, then our assumption implies the existence of $y \in A$, satisfying $x \equiv y$ (mod q). This gives

$$g(x) \equiv g(y) \equiv 0 \pmod{q},$$

and thus N divides $g(x)$, whence $f \in Int(\mathbf{Z})$.

Conversely, assume that every polynomial with rational coefficients mapping A into $\mathbf{Z}$ lies in $Int(\mathbf{Z})$, let $q = p^a$ ($a \geq 1$) be a power of a prime p and assume the existence of an integer i_0 which is not congruent mod q to any element of A.

Put $F(X) = q! h_q(X)$ (where h_q is given by (2.1)) and

$$G(X) = p^{-b-1} \frac{F(X - i_0)}{X - i_0},$$

where b is the highest power of p dividing $(q - 1)!$. In view of $F(0) = 0$, G is a polynomial. If now $x \in A$ then q does not divide $x - i_0$. Since for all $x \in \mathbf{Z}$ we have

$$p^{a+b} \mid q! \mid F(x - i_0)$$

the number p^{b+1} has to divide $\dfrac{F(x - i_0)}{x - i_0}$ showing that $G(x) \in \mathbf{Z}$ and thus G lies in $Int(\mathbf{Z})$. However

$$G(i_0) = \frac{F'(0)}{p^{b+1}} = \frac{(-1)^{q-1}(q - 1)!}{p^{b+1}} \notin \mathbf{Z},$$

a contradiction. $\square$

Actually R.Gilmer proved a more general result, showing that the analogue of Theorem 2.6 holds for all Dedekind domains R having the *finite norm property* (i.e. in which for every non-zero ideal I the factor ring R/I is finite) prime powers being replaced in this case by powers of prime ideals. Subsets A of a domain R for which the equality $Int(A, R) = Int(R)$ holds (with K being the field of quotients of R) have been called *full subsets* in P.-J.CAHEN [93], where their properties have been studied. D.L.MCQUILLAN [91] obtained a characterization of pairs of subsets S_1, S_2 of a Dedekind domain R with finite norm property which satisfy $Int(S_1, R) = Int(S_2, R)$. The corresponding question for arbitrary domains is still unanswered. (**PROBLEM II**).

6. If R is a field, then evidently one has $Int(R) = R[X]$, however Theorem 2.1 shows that in other cases $Int(R)$ may be larger than $R[X]$. In fact this happens rather often:

THEOREM 2.7. *If R is a Noetherian domain and there is a proper principal ideal of R such that all prime ideals containing it are of finite index then $Int(R) \neq R[X]$.*

PROOF: If aR is a proper principal ideal of R then with suitable prime ideals P_i we can write

$$P_1 \cdots P_r \subset aR \subset P_1 \cap \cdots \cap P_r.$$

Assume that all P_i's are of finite index. We need a simple lemma:

LEMMA 2.8. *If two ideals I, J in a Noetherian domain R are of finite index, then their product IJ also has its index finite.*

PROOF: Let $m_1, m_2, \ldots, m_t$ be generators of I as an R-module, let B, C be complete sets of representatives of residue classes $\bmod\, I$ and $\bmod\, J$ respectively. If now $x \in R$ then with a suitable $b \in B$ and $r_1, r_2, \ldots, r_t \in R$ we can write

$$x = b + \sum_{j=1}^{t} r_j m_j.$$

Choosing for each j an element $c_j \in C$ with $r_j \equiv c_j \pmod J$ we arrive at

$$x \equiv b + \sum_{j=1}^{t} c_j m_j \pmod{IJ},$$

showing that the number of residue classes mod IJ does not exceed $t \# B \cdot \# C$. $\square$

This lemma implies that the index of $P_1 \cdots P_r$ is finite and so is the index of aR. If $\alpha_1, \alpha_2, \ldots, \alpha_k$ represent all residue classes mod aR, then the polynomial

$$\frac{1}{a} \prod_{j=1}^{k} (X - \alpha_j)$$

lies in $Int(R)$ but not in $R[X]$. $\square$

7. One sees easily that $Int(R) = R[X]$ holds for a domain R if and only if for every proper principal ideal I of R only the zero polynomial can vanish identically on the factor-ring R/I. We present now some examples of such domains:

THEOREM 2.9. (i) (F.SHIBATA, T.SUGATANI, K.YOSHIDA [86]) *If R is a domain containing an infinite field, then $Int(R) = R[X]$.*

  (ii) (P.L.CAHEN, J.-L.CHABERT [71]) *If A is an infinite domain and R is the ring of polynomials in one variable over A, then $Int(R) = R[X]$.*

  (iii) (D.BRIZOLIS[76]) *If R is a subring of the ring of all algebraic integers which is closed under the operation of taking square roots, then $Int(R) = R[X]$.*

PROOF: (i) If $Int(R)$ contains properly $R[X]$ then there is a proper ideal I of R and a polynomial $f \in R[X] \setminus I[X]$, mapping R into I. The polynomial $g = f \bmod I$ maps the factor-ring R/I in 0, however R/I contains an isomorphic copy of the infinite field contained in R hence g must be the zero polynomial, a contradiction.

(ii) Let R be the ring of polynomials in one variable T over an infinite domain A and let

$$f(X) = \frac{1}{q(T)} \sum_{j=0}^{N} A_j(T) X^j \in Int(R),$$

with $q, A_0, \ldots, A_N \in R$.

Define for $j = 0, 1, \ldots, N$ the polynomials B_j by the conditions

$$B_j(T) \equiv A_j(T) \pmod{q(T)}, \quad (\deg B_j < \deg q \text{ or } B_j = 0).$$

Then the polynomial

$$\frac{\sum_{j=0}^{N} B_j(T) X^j}{q(T)}$$

maps R into R, thus for every $a \in A$

$$g_a(T) = \frac{\sum_{j=0}^{N} B_j(T) a^j}{q(T)}$$

is a polynomial over R. However the degree of its numerator is smaller than that of the denominator, and this is possible only if for all $a \in A$ we have $g_a(T) q(T) = 0$. Since A is infinite this means that all B_j's vanish, thus q divides $A_0, \ldots, A_N$ and f lies in $R[X]$ as asserted.

(iii) Let R be a subring of the ring of all algebraic integers having the property that if $a \in R$, then $\sqrt{a} \in R$. Let $F(X)$ be a polynomial with coefficients in R, let $d \in R$ be a non-unit and finally assume that the polynomial $f(X) = F(X)/d$ maps R in R.

Without restricting the generality we may assume that d does not divide the leading coefficient c of F. Denote by M the field generated by c, d and all roots of F and write

$$F(X) = c(X - a_1)(X - a_2) \cdots (X - a_m),$$

with $a_1, \ldots, a_m \in M$. For any prime ideal P of the ring of integers of M denote by v_P the multiplicative valuation of M associated with P. Since d does not divide c there exists a prime ideal P with $v_P(c) > v_P(d)$. Let v be a fixed extension of v_P to the field of all algebraic numbers and choose π in $P \setminus P^2$.

Now we show that for every positive ϵ one can find an element $u = u(\epsilon)$ in R satisfying

$$v(u - a_i) > 1 - \epsilon \quad (i = 1, 2, \ldots, m).$$

We may assume that with a suitable $t \geq 0$ we have $v(a_i) \leq 1 - \epsilon$ for

$i = 1, 2, \ldots, t$ and $v(a_i) > 1 - \epsilon$ for $i = t + 1, \ldots, m$.

If $t = m$ then we may put $u = 0$, and otherwise consider $u_r = \pi^{1/2^r}$ for $r = 1, 2, \ldots$. If $1 \le i \le t$ and $v(u_r - a_i) \le 1 - \epsilon$, then

$$v(u_r) \le \max\{v(u_r - a_i), v(a_i)\} \le 1 - \epsilon,$$

and this can hold only for finitely many values of r since $v(u_r)$ tends to 1.

In the case $t < i \le m$ if $v(a_i) \ge 1$, then in view of $v(u_r) < 1$ we obtain

$$v(u_r - a_i) = \max\{v(a_i), v(u_r)\} = v(a_i) \ge 1 > 1 - \epsilon,$$

and if $v(a_i) < 1$ then for large r we get $v(a_i) < v(u_r)$ and this gives

$$v(u_r - a_i) = \max\{v(u_r), v(a_i)\} = v(u_r),$$

thus it remains to observe that for large r one has $v(u_r) > 1 - \epsilon$.

The inequality

$$v(F(u)) \ge v(c)(1 - \epsilon)^m$$

follows and since $v(c) > v(d)$ and the factor $(1 - \epsilon)^m$ can be made arbitrarily close to unity we obtain finally $v(F(u)) > v(d)$. This shows that d does not divide $F(u)$ and $f(u) \notin R$, contradiction. $\square$

For another proof of (i) see D.D.ANDERSON, D.F.ANDERSON, M.ZAFRUL-LAH [91] and of (ii) in case of zero characteristics see G.GERBOUD [86] (Proposition 3.1).

P.-J.CAHEN and J.-L.CHABERT ([71]) gave another class of examples by showing that if R is an integrally closed domain with $Int(R) = R[X]$ and A is its integral closure in a finite extension of K, then $Int(A) = A[X]$. Still another examples have been found by R.GILMER ([90], Theorem 2), among them the ring of integers in the field of algebraic numbers generated by all roots of unity of prime order. This ring has been the first known example of a non-Noetherian almost Dedekind domain (S.NAKANO [53]). (A domain is called *almost Dedekind* provided its localizations with respect to all its prime ideals are Noetherian valuation domains). It has been shown by G.GERBOUD ([86], Theorem 3.1) that for a Dedekind domain R the condition $Int(R) = R[X]$ holds if and only if all its maximal ideals are of infinite index. A description of all domains R with $Int(R) = R[X]$ is not known (**PROBLEM III**). (See F.SHIBATA, T.SUGATANI, K.YOSHIDA [86] for the Noetherian case).

Exercises

1. Let $f(X) = \sum_{i=0}^n a_i \binom{X}{i}$ with $a_i \in \mathbf{Z}$.

(i) Prove that $f \in \mathbf{Z}[X]$ holds if and only if for $i = 2, 3, \ldots, N$ one has $i! | a_i$.

(ii) Let q be a given integer. Prove that $f \in \mathbf{Z}[X]$ and $f(\mathbf{Z}) \subset q\mathbf{Z}$ holds if and only if for $i = 0, 1, \ldots, N$ the least common multiple of q and $i!$ divides a_i.

2. (F.GRAMAIN [90]) Let $q \geq 2$ be a rational integer. Show that the set of all polynomials $f \in \mathbf{Q}[X]$ which satisfy $f(q^k) \in \mathbf{Z}$ for $k = 0, 1, 2, \ldots$ is generated as a $\mathbf{Z}$-module by the polynomials $G_0, G_1, G_2, \ldots$ defined by

$$G_0(X) = 1$$

$$G_n(X) = \frac{\prod_{j=1}^{n-1}(X - q^j)}{c_n(q)} \quad (n = 1, 2, 3, \ldots),$$

where

$$c_n(q) = q^{n(n-1)/2} \prod_{j=1}^{n}(q^j - 1).$$

3. (P.-J.CAHEN, J.-L.CHABERT [71]) Let R be a commutative ring and A an R-module. A is said to be *without polynomial torsion*, (WPT) if for every polynomial

$$f(X) = \sum_{j=0}^{N} a_j X^j$$

with $a_j \in A$ from $f(R) = 0$ the equalities $a_0 = a_1 = \cdots = a_N = 0$ follow.

(i) Show that if A_i is a family of WPT-modules, then their direct sum and direct product also are WPT,

(ii) Show that a submodule of a WPT-module is also WPT,

(iii) Show that if A is an R-module, B its submodule and both modules B and A/B are WPT, then so is A.

(iv) Show that if A is an R-module which is WPT, then for every polynomial $f \in A[X_1, \ldots, X_n]$ the condition $f(R^n) = 0$ implies $f = 0$.

4. (P.-J.CAHEN, J.-L.CHABERT [71])
(i) Prove that an R-module A is WPT if and only if for every non-zero $a \in A$ the ring $M = R/An(x)$ (where $An(x) = \{r \in R : rx = 0\}$ is the *annihilator* of x) is WPT as an M-module.

(ii) Let R be a Noetherian ring. Show that if A is an R-module then it is WPT if and only if every its associated prime ideal is of infinite index in R.

(iii) Give an example to show that the assertion in (ii) may fail for non-Noetherian rings.

5. (P.-J.CAHEN, J.-L.CHABERT[71])
(i) Let R be a commutative infinite ring and let A be an R-module. Prove that if $f \in A[X]$ vanishes at all except finitely many elements of R, then it vanishes at all $x \in R$.

(ii) Let R be a domain and K its quotient field. Show that if $f \in K[X]$ and for all except finitely many elements $r \in R$ one has $f(r) \in R$, then $f \in Int(R)$.

6. (P.-J.CAHEN, J.-L.CHABERT [71]) Let R be a domain. Prove that an R-module A is WPT if and only if A is infinite and the torsion-submodule of A is WPT.

7. (J.ACZEL [60]) Let R be a domain such that $nr = 0$ for $n \in \mathbf{Z}$ and $r \in R$ implies $n = 0$ or $r = 0$. Prove that R is WPT if treated as an R-module.

III. Fixed divisors

1. In the next section we shall deal with the analogue of Theorem 2.1 for certain algebraic number fields. To do that we need a result of G.PÓLYA [19] concerning fixed ideal divisors of polynomials. We start with the definition of a fixed divisor:

Let f be a polynomial with coefficients in a domain R and assume that the coefficients of f generate the unit ideal. If f maps R into a proper ideal I of R, then I is called a *fixed divisor* of f.

The following result (due to G.PÓLYA [19] in case when R is the ring of integers of an algebraic number field) describes prime ideal powers in a class of Dedekind domains which can be fixed divisors of a suitable polynomial of a given degree.

THEOREM 3.1. *Let R be a Dedekind domain with the finite norm property. Let P be a prime ideal of R, let $N(P) = \#R/P$ be its norm and let f be a polynomial of degree n over R not all coefficients of which lie in P. If f maps R into P^a for some positive a, then*

$$(3.1) \qquad a \leq A(n, N(P)),$$

where

$$A(n,k) = \sum_{j \geq 1} \left[\frac{n}{k^j} \right] < \frac{n}{k-1} \leq n.$$

Moreover for every n there exists a monic polynomial f of degree n for which equality occurs in (3.1).

PROOF: Put $N = N(P)$ and let $a_0 = 0, a_1, \ldots, a_{N-1}$ be a complete residue system mod P. Choose π in $P \setminus P^2$ and let m be a nonnegative integer. If

$$m = c_0 + c_1 N + \cdots + c_h N^h$$

with $0 \leq c_j < N$, then define

$$a_m = a_{c_0} + a_{c_1} \pi + \cdots + a_{c_h} \pi^h.$$

Note that for $m < N$ this agrees with the definition of a_i given previously for $i = 0, 1, \ldots, N - 1$.

LEMMA 3.2. *For any integer $k \geq 1$ one has $\pi^k \| a_m - a_n$ if and only if N^k divides $m - n$ but N^{k+1} does not.*

PROOF: If

$$m = \sum_{j \geq 0} c_j N^j, \quad n = \sum_{j \geq 0} d_j N^j,$$

with $0 \leq c_j, d_j < N$ then

$$m - n = (c_r - d_r)N^r + \cdots,$$

where r is the smallest index for which $c_r \neq d_r$. On the other hand we have

$$a_m - a_n = (a_{c_r} - a_{d_r})\pi^r + \cdots$$

and the assertion results immediately. $\square$

COROLLARY. For $k = 1, 2, \ldots$ the numbers $a_0, a_1, \ldots, a_{N^k-1}$ form a complete residue system mod P^k. $\square$

Now put $f_0(X) = 1$ and

(3.2)
$$f_m(X) = \prod_{j=0}^{m-1} (X - a_j)$$

for $m = 1, 2, \ldots$.

LEMMA 3.3. For $m = 1, 2, \ldots$ one has $P^{A(m,N)} \| f_m(a_m)$.

PROOF: If for any integer a we denote by $w_N(a)$ the exponent of the maximal power of N dividing a then the preceding lemma shows that the exponent of the maximal power of P which divides

$$f_m(a_m) = (a_m - a_0) \cdots (a_m - a_{m-1})$$

equals

$$\sum_{j=0}^{m-1} w_N(m - j) = \sum_{j=1}^{m} w_N(j).$$

Observe now that for $k = 1, 2, \ldots$ the number of integers $1 \leq j \leq m$ with $w_N(j) = k$ equals $[m/N^k] - [m/N^{k+1}]$ and this gives

$$\sum_{j=1}^{m} w_N(j) = \left[\frac{m}{N}\right] - \left[\frac{m}{N^2}\right] + 2\left(\left[\frac{m}{N^2}\right] - \left[\frac{m}{N^3}\right]\right) + \cdots = A(m, N). \ \square$$

COROLLARY. For $m = 1, 2, \ldots$ the values attained by $f_m(X)$ on R are all divisible by $P^{A(m,N)}$, but not by a higher power of P.

PROOF: The second assertion is an immediate consequence of the Lemma. To prove the first observe that by Corollary to Lemma 3.2 every element x of R is congruent mod $P^{A(m,N)}$ to some a_j, with $0 \leq j \leq N^{A(m,N)} - 1$. Thus

$$f_m(x) \equiv f_m(a_j) \equiv (a_j - a_0) \cdots (a_j - a_{m-1}) \pmod{P^{A(m,N)}},$$

and so our assertion holds in case $j \leq m - 1$. If however j exceeds $m - 1$, then as in the proof of the lemma we obtain

$$\sum_{i=0}^{m-1} w_N(j-i)$$

$$= \sum_{i \geq 1} i \left(\left(\left[\frac{j}{N^i} \right] - \left[\frac{j-m}{N^i} \right] \right) - \left(\left[\frac{j}{N^{i+1}} \right] - \left[\frac{j-m}{N^{i+1}} \right] \right) \right)$$

$$= \sum_{i \geq 1} \left(\left[\frac{j}{N^i} \right] - \left[\frac{j-m}{N^i} \right] \right) \geq \sum_{i \geq 1} \left[\frac{m}{N^i} \right] = A(m, N). \quad \square$$

Now we can conclude the proof of the theorem. We deal only with its first assertion, since the second is contained in the preceding corollary. Assume that P is a prime ideal of R and $F(X)$ is a polynomial of degree n over R whose not all coefficients are divisible by P and which maps R in P^B, where $B = A + 1$, $A = A(n, N(P))$. Since all polynomials f_j are unitary we can write $F(X)$ in the form

$$(3.3) \qquad\qquad F(X) = \sum_{j=0}^{n} c_j f_j(X)$$

with suitable $c_j \in R$. We shall show that all c_j's must be divisible by P. Assume, to the contrary, that for a certain $0 \leq s \leq n$ the number c_s is not divisible by P and denote by t the smallest index with the property

$$c_t \not\equiv 0 \pmod{P^{B-A(t,N(P))}}.$$

The set of such indices is non-empty since it contains s. Indeed, we have

$$B - A(s, N(P)) = B - A + A - A(s, N(P)) = 1 + A(n, N(P)) - A(s, N(P)) > 0.$$

Putting now in (3.3) $X = a_t$ and using the Corollary to Lemma 3.3 we arrive at

$$0 \equiv F(a_t) \equiv \sum_{j=0}^{n} c_j f_j(a_t) \equiv \sum_{j=0}^{t} c_j f_j(a_t) \equiv c_t f_t(a_t) \pmod{P^B},$$

because for $j > t$ we have $f_j(a_t) = 0$ and for $j < t$ we have

$$c_j \equiv 0 \pmod{P^{B-A(j,N(P))}}$$

and

$$f_j(a_t) \equiv 0 \pmod{P^{A(j,N(P))}}.$$

Since $f_t(a_t)$, according to Lemma 3.3, is not divisible by $P^{A(t,N(P))+1}$ it follows that c_t must be divisible by $P^{B-A(t,N(P))}$, a contradiction. Thus all c_j's are divisible by P and (3.3) shows now that the same applies to all coefficients of F. $\quad \square$

2. It has been observed by P.J.CAHEN [72] that the assertion of Theorem 3.1 holds also for Dedekind domains R in which there are prime ideals P with infinite factor-rings R/P provided for such P we put $A(k, N(P^i)) = 0$ for $i = 1, 2, \ldots$.

COROLLARY 1. *Let R be a Dedekind domain with finite norm property. If*

$$f(X) = \frac{c_k X^k + \cdots + c_0}{q}$$

is a polynomial over K (with $c_0, c_1, \ldots, c_k, q \in R$ and $\sum_{j=1}^{k} c_j R + qR = R$) which maps R in R, P is a prime ideal of R and $\nu_P(x)$ is the exponent associated with P, then

$$\nu_P(q) \le A(k, N(P)).$$

PROOF: It suffices to observe that the polynomial $qf(X)$ maps R into the ideal generated by q. □

COROLLARY 2. *If R is a Dedekind domain with finite norm property, I is its proper ideal, and $m > 1$, then I is a fixed divisor of a suitable monic polynomial of degree m over R if and only if I divides the ideal*

$$\mathcal{A} = \prod_P P^{A(m, N(P))}.$$

PROOF: The necessity is contained in Theorem 3.1. To prove the sufficiency, for every prime ideal P dividing $\mathcal{A}$ denote by $f_{m,P}$ the polynomial defined by (3.2) and write A_P for $A(m, N(P))$. By the Chinese Remainder Theorem there exists a polynomial $f \in R[X]$ satisfying the congruences

$$f \equiv f_{m,P} \pmod{P^{A_P}} \quad (P|\mathcal{A}),$$

and obviously it maps R into $\mathcal{A}$. □

In H.GUNJI,D.L.MCQUILLAN [70] an analogue of Corollary 2 has been obtained for the common divisor of values attained by a polynomial on a fixed coset relative to a non-zero ideal.

3. A simple way of determining all fixed principal divisors of a given polynomial in arbitrary many variables over $\mathbf{Z}$ has been given by K.HENSEL [96]. We prove now his result in the case of one variable:

THEOREM 3.4. *If $f \in \mathbf{Z}[X]$ is a polynomial of degree n and $d \in \mathbf{Z}$ then one has $f(\mathbf{Z}) \subset d\mathbf{Z}$ if and only if d divides the numbers $f(i)$ for $i = 0, 1, \ldots, n$.*

PROOF: The assertion being obvious in case $n = 0$ assume that it holds for all polynomials of degree $n - 1$ and let $f \in \mathbf{Z}[X]$ be of degree n. Put $g(X) = f(X + 1) - f(X)$. If the values $f(0), f(1), \ldots, f(n)$ are divisible by d, then d divides $g(i)$ for $i = 0, 1, \ldots, n - 1$ and since deg $g = n - 1$ we get $g(\mathbf{Z}) \subset d\mathbf{Z}$. This implies

$$f(i + 1) \equiv f(i) \pmod{d} \quad (i = 0, 1, \ldots),$$

and $f(\mathbf{Z}) \subset d\mathbf{Z}$ results. $\square$

If R is the ring of integers either of a global or a local field and P a prime ideal of R then it has been shown by D.J.LEWIS [56] that the ideal of $R[X]$ consisting of all polynomials f with $f(R) \subset P^m$ is generated by at most $1 + m$ elements, and gave an explicit construction of a set of generators.

Exercises

1. Let R be a Dedekind domain with the finite norm property. Show that if R is not a unique factorization domain, then there are polynomials in $R[X]$ having fixed ideal divisors but no principal fixed divisors.

2. (K.HENSEL [96]) Generalize Theorem 3.4 to the case of several variables.

3. (J.TURK [86]) For a polynomial $f \in \mathbf{Z}[X]$ denote by $H(f)$ its *height* (i.e. the maximal modulus of its coefficients) and let $a(d, h)$ be the number of $f \in \mathbf{Z}[X]$ which satisfy $\deg f \le d$ and $H(f) \le h$. Moreover denote by $b(d, h, N)$ the number of such f which additionally satisfy $f(\mathbf{Z}) \subset N\mathbf{Z}$.

(i) Prove that

$$\lim_{h \to \infty} \frac{b(d, h, N)}{a(d, h)} = \prod_{j=0}^{d} \frac{(j!, N)}{N}.$$

(ii) Show that if $c(d, h, N)$ denotes the number of $f \in \mathbf{Z}[X]$ with $\deg f \le d$ and $H(f) \le h$ which have the maximal fixed divisor equal to N then

$$\lim_{h \to \infty} \frac{c(d, h, N)}{a(d, h)} = \sum_{i=0}^{\infty} \mu(i) \prod_{j=0}^{d} \frac{(j!, N)}{iN},$$

with $\mu(i)$ being the Möbius function.

(iii) Prove that the "probability" for a polynomial in $\mathbf{Z}[X]$ to have no fixed divisor > 1 equals

$$\lambda = \prod_{p} (1 - p^{-p}),$$

i.e.

$$\lim_{d \to \infty} \lim_{h \to \infty} \frac{c(d, h, 1)}{a(d, h)} = \lambda.$$

(iv) Obtain the analogue of (i)-(iii) for polynomials in several variables.

IV. Regular basis

1. Theorem 2.1 shows that in the case $R = \mathbf{Z}$ the ring $Int(R)$ is a free R-module generated by a sequence of polynomials $f_0, f_1, \ldots$ with $\deg f_i = i$ for $i = 0, 1, 2, \ldots$ and with the property that for $k = 1, 2, \ldots$ every polynomial of degree k belonging to $Int(R)$ is uniquely expressible as a linear combination of $f_0, f_1, \ldots, f_k$ with coefficients from R. G.PÓLYA [19] asked whether the same result holds for rings of all algebraic integers in an algebraic number field, proved it in the case when this ring is a unique factorization domain, noted that it may fail in general and gave a necessary and sufficient condition for this property to hold in a quadratic number field. Such condition has been given for all algebraic number fields by A.OSTROWSKI [19]. Rings with this property are said to have a *regular basis* and algebraic number fields whose rings of integers have a regular basis are called *Pólya fields*.

2. Before presenting these results we prove a simple theorem, essentially due to G.PÓLYA ([19], Satz III) , which gives information about the R-module structure of $Int(R)$ and forms a weaker form of Theorem 2.1 for Dedekind domains. Pólya formulated it only for rings of algebraic integers but his idea works without essential changes in the general case as observed by P.J.CAHEN[72]. To avoid certain technical inconveniences we restrict ourselves to the case, when R has the finite norm property and has zero characteristic.

First we define a sequence of ideals in an arbitrary domain R, which will be useful also at a later stage. Denote for $m = 0, 1, 2, \ldots$ by I_m the set of all leading coefficients of members of $Int(R)$ of degree m.

LEMMA 4.1. *Let R be a Dedekind domain of zero characteristic and the finite norm property. Then the I_m's are fractional ideals of R and moreover $J_m = m! I_m$ are ideals of R. Moreover for $m = 1, 2, \ldots$ we have $I_m = \prod P^{a_P}$, (where P runs over all prime ideals of R) then $a_P = -A(m, N(P))$, $A(m, k)$ being given by Theorem 3.1. Moreover I_m^{-1} is an integral ideal of R coinciding with the ideal $\mathcal{A}$ occuring in Corollary 2 to Theorem 3.1.*

PROOF: Put $A = A(m, N(P))$, let π be an element of $P \setminus P^2$ and define $f_m \in R[X]$ by (3.2). Write $\pi R = PI$, where I is an ideal not divisible by P and observe that if $c \in I \setminus P$ then the polynomial

$$g(X) = \frac{c^A f_m(X)}{\pi^A}$$

maps R into R. Indeed, write $cR = IJ$, where J is an ideal of R and observe that in view of Corollary to Lemma 3.3 for every x in R the fractional ideal generated by $g(x)$ is contained in J^A and hence $g(x) \in R$. Invoking again that corollary we get $a_P \leq -A$.

To obtain the converse inequality assume it to be false and let g be a polynomial over K of degree k which maps R into R with leading term c satisfying $n_P(c) \leq -A - 1$, where n_P is the exponent associated with P. If q denotes the common denominator of all coefficients of g, then P^{1+A} divides qR and so the polynomial $qg(X)$ maps R into P^{1+A} which in view of Theorem 3.1 is not possible. The second assertion follows now immediately. $\square$

THEOREM 4.2. *Let R be a Dedekind domain of zero characteristic and the finite norm property.*

(i) *The R-module $Int(R)$ is isomorphic to the direct sum*

$$\bigoplus_{m=0}^{\infty} I_m,$$

and R has a regular basis if and only if all ideals I_m are principal.

(ii) *There exists a sequence $f_0, f_1, \ldots$ of polynomials in $Int(R)$ with $\deg f_i = i$ for $i = 0, 1, 2, \ldots$ and a set $M = \{m_1 < m_2 < \ldots\}$ of positive integers (which may be empty) such that every element f of $Int(R)$ can be represented in the form*

$$f = \sum_{j=0}^{N} c_j f_j + \sum_{i=1}^{n} d_i x^{m_i},$$

with $N = \deg f, n \geq 0, m_i \leq N$ and $c_j, d_i \in R$. An integer m belongs to M if and only if the ideal J_m is non-principal. This representation of f is unique if and only if M is empty.

PROOF: (i) Using the Chinese Residue Theorem one gets from Theorem 3.1 and Lemma 4.1 the existence of monic polynomials $F_0 = 1, F_1, F_2, \ldots$ in $R[X]$ satisfying $\deg F_n = n$ and $F_n(R) = I_n^{-1}$ for $n = 1, 2, \ldots$. Every polynomial $f \in K[X]$ can be uniquely written in the form $f = \sum_{j=0}^{N} \alpha_j F_j$ with $N = \deg f$ and $\alpha_j \in K$. To establish the first part of the theorem it suffices to show that f belongs to $Int(R)$ if and only one has $\alpha_j \in I_j$ for $j = 0, 1, \ldots$. To do this observe first that if for $j = 0, 1, \ldots, N$ we have $\alpha_j \in I_j$, then $\alpha_j F_j(R) \subset I_j F_j(R) \subset R$, hence $\alpha_j F_j \in Int(R)$ and $f \in Int(R)$ follows. Conversely, if $f \in Int(R)$, then α_N equals the leading coefficient of f and thus lies in I_N, but the previous argument gives $\alpha_N F_N \in Int(R)$, and ths leads to $\sum_{j=0}^{N-1} \alpha_j F_j = f - \alpha_N F_N \in Int(R)$ and by recurrence we get $\alpha_j \in Int(R)$ for all j.

If all ideals I_m are principal, generated by a_m, say, then for each m choose a polynomial f_m of degree m in $IntR$ having a_m for its leading coefficients. Clearly the sequence $f_0, f_1, \ldots$ forms a regular basis.

Conversely, if $f_0, f_1, \ldots$ is a regular basis and $a \in J_m$, then there exists a polynomial $f \in Int(R)$ with

$$f(X) = \frac{a}{m!} X^m + \cdots .$$

Define $c \in R$ by

$$f_m(X) = \frac{c}{m!} X^m + \cdots$$

and observe that from the equality

$$f = c_m f_m + \cdots + c_0 f_0$$

one gets $a = c c_m$, and we obtain that J_m is principal, generated by c.

(ii) Let M be the set of all such integers m for which the ideal I_m is not principal. If $m \in M$ then J_m is also non-principal and because the polynomial x^m lies in $Int(R)$ we get $1 \in I_m$ and $m! \in J_m$. Since R is a Dedekind domain there exists an element $a \in J_m$ such that $J_m = m!R + aR$. For such m's put $f_m = f_{m,a}$, where $f_{m,a}$ is defined as in part (i). In case of the remaining m's the ideal J_m is principal and we put $f_m = f_{m,b}$, where b is any generator of J_m.

If M is empty, then (i) implies the uniqueness of representation. If however M contains an element m, then J_m is not principal and if we write

$$f_m = \frac{a}{m!} X^m + \cdots$$

then the polynomial $m! f_m - aX^m$ lies in $Int(R)$, is of degree $< m$ and thus can be expressed as a linear combination of $f_0, \ldots, f_{m-1}$ and of certain powers x^k with $k \le m - 1$. This shows that it has at least two different representations. $\square$

COROLLARY. *If R is a principal ideal domain then it has a regular basis.*

PROOF: In this case the set M in (ii) is empty. $\square$

If R is a Dedekind domain then all its non-zero ideals are projective, hence $Int(R)$ is a projective R-module, being a direct sum of projectives modules. It follows from a result of H.BASS [63] (see also [B], sect.4, Exercises) that $Int(R)$ is a free R-module and hence there exists a sequence $f_0, f_1, \ldots$ of elements of $Int(R)$ with the property that every element of $Int(R)$ can be uniquely represented as a linear combination of the f_i's with coefficients from R. Note however that in general the condition deg $f_m = m$ may not be satisfied.

An analogue of Theorem 4.2 for $Int(I, J)$ where I, J are ideals in a Dedekind domain having the finite-norm property has been considered in D.L.McQUILLAN [73a],[73b] and H.GUNJI, D.L.McQUILLAN [78]. Polynomials mapping R into a fixed principal ideal were considered by G.JACOB ([76],[80]) who gave an explicit direct decomposition of the free R-module formed by these polynomials and in the case of a principal ideal domain found an explicit basis of it. A similar result has been in case $R = \mathbf{Z}$ obtained by D.SINGMASTER [74], who showed that every polynomial $f \in \mathbf{Z}[X]$ mapping $\mathbf{Z}$ into $m\mathbf{Z}$ can be written in the form

$$f(X) = g_1(X)S_n(X) + \sum_{k=0}^{n-1} a_k \frac{m}{(k!,m)} S_k(X) + mg_2(X),$$

where g_1, g_2 are $\mathbf{Z}$-polynomials, $n = n(m)$ denotes the minimal integer satisfying $m \mid n!$, $S_0(X) = 1$, $S_k(X) = (X + 1) \cdots (X + k)$ $(k = 1, 2, \ldots)$ and $a_0, a_1, \ldots, a_{n-1}$ are integers. Conversely, every polynomial of this form maps $\mathbf{Z}$ into $m\mathbf{Z}$.

Exercises

1. Show that if a domain R has a regular basis $g_0, g_1, \ldots$ then every polynomial in n variables mapping R^n in R is of the form

$$f(X_1, \ldots, X_n) = \sum_{i_1, \ldots, i_n} c(i_1, \ldots, i_n) g_{i_1}(X_1) \cdots g_{i_n}(X_n),$$

with $c(i_1, \ldots, i_n) \in R$.

2. (G.JACOB [76]) Let R be a domain having K for its field of quotients. Assume that there exists a sequence of polynomials $f_n \in Int(R)$ with $\deg f_n = n$ and a sequence of R-modules $I_n \subset K$ such that

$$Int(R) = \bigoplus_{n=0}^{\infty} I_n f_n.$$

(i) Show that if $I = aR$ is a principal ideal of R, then the ideal of $R[X]$ consisting of all polynomials mapping R in I equals

$$\bigoplus_{n=0}^{\infty} J_n f_n,$$

where $J_n = (aI_n) \cap R$.

(ii) Determine the ideals J_n in case when R is a unique factorization domain.

3. (G.JACOB [76]) Let R be a Dedekind domain with a regular basis and for any ideal $I \subset R[X]$ put

$$A_I(R) = \{f \in R[X] : f(R) \subset I\}.$$

(i) Show that if I, J are coprime ideals of R, then

$$A_{IJ}(R) = A_I(R) \oplus A_J(R).$$

(ii) Describe $A_q(R)$ for a prime ideal power q.

V. Pólya fields

1. The following theorem of A.OSTROWSKI [19] characterizes Pólya fields:

THEOREM 5.1. *If K is an algebraic number field and R its ring of integers then the following conditions are equivalent:*

(i) *K is a Pólya field,*

(ii) *For every prime power p^s the product $b(p^s)$ of all prime ideals of R with norm p^s is principal.*

PROOF: (i) $\Longrightarrow$ (ii). Observe first that if q is a given prime-power and I_q denotes the ideal defined in the preceding section, then the Lemma 4.1 gives

$$I_q = \prod_{p,s} b(p^s)^{-A(q,p^s)} = \prod_{\substack{p,s \\ p^s \le q}} b(p^s)^{-A(q,p^s)} = b(q) \prod_{\substack{p,s \\ p^s < q}} b(p^s)^{-A(q,p^s)},$$

and since our assumption implies (in view of Theorem 4.2) that all ideals I_q are principal, we obtain by recurrence the principality of all ideals $b(q)$.

(ii) $\Longrightarrow$ (i). Considering again the equality

$$I_q = \prod_{p,s} b(p^s)^{-A(q,p^s)},$$

we see that (ii) implies the principality of I_q and again Theorem 4.2 leads to the assertion. $\square$

It is now easy to obtain a simple condition for a normal field K to be a Pólya field:

COROLLARY 1. *If $K/\mathbf{Q}$ is a finite normal extension with Galois group G, then the following conditions are equivalent:*

(i) *K is a Pólya field,*

(ii) *For every prime p ramified in $K/\mathbf{Q}$ the product of all prime ideals of K lying over p is principal,*

(iii) *Every fractional ideal of K invariant under the action of G is principal.*

PROOF: (i) $\Longleftrightarrow$ (ii) Since the extension $K/\mathbf{Q}$ is normal we have for every rational prime p with a suitable e the equality

$$pR = (P_1 \cdots P_g)^e,$$

where $P_1, \ldots, P_g$ are prime ideals of the same norm, say p^f. Then $b(p^a)$ equals $P_1 \cdots P_g$ in case $a = f$ and equals the unit ideal otherwise. If p is unramified, then $e = 1$, thus $b(p^f) = pR$ and so it suffices to consider only ramified primes in which case the assertion follows immediately from the theorem.

(ii) $\Longleftrightarrow$ (iii) It suffices to observe that the group of invariant fractional ideals of K is generated by the set $\{b(q)\}$ where q runs over all prime powers. $\square$

COROLLARY 2. (i) (G.PÓLYA [19]) *A quadratic number field K is a Pólya field if and only if all ramified prime ideals of K are principal.*

(ii) (H.ZANTEMA [82]) *Every cyclotomic field is a Pólya field.*

PROOF: (i) Recall that the factorization law of primes in a quadratic field implies that every ramified prime p is of the first degree and in this case $b(p)$ is a prime ideal, hence it suffices to apply the preceding corollary.

(ii) Let $K = \mathbf{Q}(\zeta_m)$ where ζ_m is a primitive m-th root of unity. We may assume that m is either odd or divisible by 4. We shall use the factorization law in cyclotomic extensions (see e.g. [EATAN], Theorem 4.16). Let p be a ramified prime. Then p divides m and if $m = p^k n$ with $p \nmid n$, ζ is a primitive root of unity of order p^k, $L = \mathbf{Q}(\zeta)$, and $\mathbf{Z}_K$, $\mathbf{Z}_L$ denote the rings of integers in K and L, respectively, then one has $p\mathbf{Z}_L = P^{[L:Q]}$ where $P = (1 - \zeta)\mathbf{Z}_L$ is a principal prime ideal. Moreover $P\mathbf{Z}_K$ is a product of distinct prime ideals, hence $b(p) = P\mathbf{Z}_K = (1 - \zeta)\mathbf{Z}_K$ is principal. $\square$

(In case of cyclotomic fields of prime power order the assertion (ii) has been established by A.OSTROWSKI [19]).

Part (i) of this Corollary allows us to give examples of non-Pólya fields. Indeed, a quadratic field K with discriminant having at least three prime divisors cannot be a Pólya field since from the usual proof of the Gauss's theorem on generas (see e.g. [EATAN], Theorem 8.8) it follows that in this case the ramified prime ideals lie in distinct generas and thus at least one of these ideals is non-principal.

The same Corollary shows that the condition $h(K) = 1$ is only sufficient but not necessary for a field K to be a Pólya field. In fact, this follows from (ii) since there are only finitely many cyclotomic fields with class-number 1 (K.UCHIDA [71], see also J.M.MASLEY, H.L.MONTGOMERY [76]), but one can argue also elementarily as follows: let $c = \sqrt{-23}$ and $K = \mathbf{Q}(c)$. Here the only ramified prime ideal is generated by c, thus is principal, so K is a Pólya field, however $\mathbf{Z}_K$ is not a unique factorization domain, as

$$6 = 2 \cdot 3 = \frac{1 + c}{2} \cdot \frac{1 - c}{2},$$

and since the equations $x^2 + 23y^2 = a$ with $a = 8, 12$ have no solutions in $\mathbf{Z}$, the elements $2, 3$ and $(1 \pm c)/2$ are irreducible in $\mathbf{Z}_K$. (Actually the class-number of K equals 3).

2. If the class-number of K equals 1 then it is not difficult to give a procedure leading to a regular basis:

Put $F_0(X) = 1$, $F_1(X) = X$ and for $k > 1$ proceed as follows: denote by $\{P_1, \ldots, P_s\}$ the set of all prime ideals with norms not exceeding k, for $i = 1, 2, \ldots, s$ put $v_i = A(k, N(P_i))$ and let π_i be a fixed generator of P_i. Finally let $f_i(X)$ be a monic polynomial over R having degree k and mapping R into $P_i^{v_i}$. The existence of such polynomials has been established in Corollary to Lemma 3.3.

Put for $i = 1, 2, \ldots, s$: $w_i = k - v_i$, $d_i = \pi_i^{v_i}$, $d = d_1 \cdots d_s$, $\vartheta_i = d/d_i$ and define

$$g_i(X) = \frac{X^{w_i} f_i(X)}{d_i}.$$

If now $t_1, \ldots, t_s$ are elements of R satisfying

$$\sum_{i=1}^{s} t_i \vartheta_i = 1,$$

then the polynomial

$$F_k(X) = \sum_{i=1}^{s} t_i g_i(X)$$

is of degree k, maps R in R and its leading term equals $1/d$, hence generates I_k. This implies (as in the proof of Theorem 4.2 (i)) that the sequence $F_0, F_1, \ldots$ forms a regular basis.

A procedure in the general case has been given by G.GERBOUD [89], who treated arbitrary Dedekind domains. (For the particular cases $K = \mathbf{Q}(i)$, $\mathbf{Q}(\zeta_3)$ see G.GERBOUD [86], [88b]).

3. H.ZANTEMA [82] showed that if the extension $K/\mathbf{Q}$ is normal with Galois group G, then the cohomology group $H^1(G, U)$, where $U = U(K)$ is the group of units of K, is related to the question whether K is a Pólya field. To state this relation we have to use a few elementary facts from cohomological algebra. We shall use only the groups H^0 and H^1 and start with recalling their definitions:

If A is an abelian group on which the group G acts as a group of endomorphisms (i.e. A is a G-module) then the groups $H^0(G, A)$ and $H^1(G, A)$ are defined in the following way:

The group $H^0(G, A)$ equals A^G, the subgroup of A consisting of all elements invariant under G and the group $H^1(G, A)$ is the factor group of the group of crossed homomorphisms of G in A (i.e. maps $f : G \to A$, satisfying the equality $f(gh) = gf(h) + f(g)$ for all $g, h \in G$), by the group of principal crossed homomorphisms (i.e. maps f_a defined by $f_a(g) = ga - a$ with some fixed $a \in A$).

The properties of H^0 and H^1 which we need are collected in the following two lemmas:

LEMMA 5.2. (See e.g. [CF], Chapter 4, sections 1 and 2).

If a sequence

$$1 \longrightarrow A \overset{f}{\longrightarrow} B \overset{g}{\longrightarrow} C \longrightarrow 1$$

of multiplicatively written G-modules is exact (i.e. f is an embedding of A in B, g maps B onto C, and the kernel of g coincides with the image of f), then it induces the following exact sequence of abelian groups

$$1 \longrightarrow H^0(G,A) \longrightarrow H^0(G,B) \longrightarrow H^0(G,C) \longrightarrow H^1(G,A) \longrightarrow H^1(G,B). \quad \square$$

LEMMA 5.3. *Let $K/\mathbf{Q}$ be a finite normal extension with Galois group G, let K^* be the multiplicative group of K and let U be the group of units of K, G acting on K^* and U in the natural way.*

(i) (Hilbert's Theorem 90) $H^1(G, K^*) = 1$,

(ii) The number of summands in any non-trivial decomposition of $H^1(G, U)$ into cyclic summands is bounded by a number which depends only on the degree of the extension $K/\mathbf{Q}$.

PROOF: For (i) see [CF], Chapter 5, Proposition 1.2, and (ii) results immediately from the inequality (4) of [CF], Chapter 9, section 3. $\square$

THEOREM 5.4. *Let $K/\mathbf{Q}$ be a finite Galois extension with Galois group G, $\mathbf{Z}_K$ the ring of integers of K and U its group of units. If $p_1, \ldots, p_s$ are all primes ramified in $K/\mathbf{Q}$ and e_i is the ramification index of p_i, then there is a canonical embedding T of $H^1(G, U)$ into the direct sum C of cyclic groups of orders $e_1, \ldots, e_s$. This embedding is surjective if and only if K is a Pólya field.*

PROOF: For every rational prime p denote by $q(p)$ the norm of prime ideals of $\mathbf{Z}_K$ lying over p. This number is the only power p^u with nontrivial $b(p^u)$. It follows that the set $\{b(q(p)) : p \text{ prime}\}$ generates $I^G(K)$, the group of all fractional ideals of K invariant under G. This allows us to define a map $F : I^G(K) \longrightarrow C$ by putting

$$F(b(q(p)) = [1 \bmod e_1, \ldots, 1 \bmod e_s]$$

and extending F to $I^G(K)$ by additivity. The surjectivity of F is evident and to find its kernel observe that

$$F(\prod_p b(q(p))^{k(p)}) = 0$$

is equivalent to the divisibility of $k(p_i)$ by e_i for $i = 1, 2, \ldots, s$. Since $p_i \mathbf{Z}_K = b(q(p))^{e_i}$ $(i = 1, 2, \ldots s,)$ and $p\mathbf{Z}_K = b(q(p))$ for remaining primes p, we obtain

$$\ker F = \{a\mathbf{Z}_K : a \in \mathbf{Q}\} \simeq \mathbf{Q}^*/\{1, -1\},$$

and hence the following exact sequence results:

(5.1) $$1 \longrightarrow \mathbf{Q}^*/\{1, -1\} \longrightarrow I^G(K) \longrightarrow C \longrightarrow 1$$

We consider also another sequence, which is obviously exact:

$$1 \longrightarrow U \longrightarrow K^* \longrightarrow P(K) \longrightarrow 1,$$

where $P(K)$ denotes the group of all principal fractional ideals of K and using Lemma 5.2 we get the exact sequence

$$1 \longrightarrow H^0(G,U) \longrightarrow H^0(G,K^*) \longrightarrow H^0(G,P(K))$$
$$\longrightarrow H^1(G,U) \longrightarrow H^1(G,K^*).$$

Using again Lemma 5.2 and Lemma 5.3 (i) we arrive at

$$1 \longrightarrow \{1,-1\} \longrightarrow \mathbf{Q}^* \xrightarrow{f} P(K)^G \xrightarrow{g} H^1(G,U) \longrightarrow 1.$$

The exactness of this sequence implies

$$\ker\ g \simeq f(\mathbf{Q}^*) \simeq \mathbf{Q}^*/\{1,-1\}$$

and

$$H^1(G,U) \simeq P(K)^G/\ker\ g \simeq P(K)^G/(\mathbf{Q}^*/\{1,-1\}).$$

Using this isomorphism and (5.1) we obtain now a well-defined injection

$$T : H^1(G,U) \longrightarrow C$$

and this leads to the exact sequence

$$1 \longrightarrow H^1(G,U) \xrightarrow{T} C \longrightarrow I(K)^G/P(K)^G \longrightarrow 1.$$

The assertion follows now from the observation that T is surjective if and only if $I(K)^G = P(K)^G$, and this condition coincides with the condition (iii) of Corollary 1 to Theorem 5.1. $\square$

COROLLARY 1. *A normal field K with Galois group G is a Pólya field if and only if $\#H^1(G,U) = e_1 \cdots e_s$, where the e_i's are defined in the theorem.* $\square$

COROLLARY 2. *If K is a normal Pólya field of degree N, then the number of prime divisors of the discriminant $d(K)$ is bounded by a number depending only on N.*

PROOF: Denote by t the number of prime divisors of $d(K)$. Since these primes coincide with primes ramified in $K/\mathbf{Q}$ and their ramification indices exceed 1, the theorem implies the existence of a decomposition of $H^1(G,U)$ into a product of t cyclic factors, thus our assertion follows immediately from Lemma 5.3 (ii). $\square$

Using a more precise form of this corollary H.ZANTEMA [82] described cyclic Pólya fields:

If $K/\mathbf{Q}$ is cyclic of a prime-power degree p^a, then K is a Pólya field if either only one prime ramifies in K or $p = 2$, K is real, every unit of K has norm 1 and exactly two primes ramify in K, one of them with ramification index 2.

If $L/\mathbf{Q}$ is cyclic of a composite degree $N = \prod_p p^{a_p}$, then L is a Pólya field if and only if for all primes p dividing N the unique subfield of L of degree p^{a_p} is a Pólya field.

In the same paper it is shown that if the Galois group of the Galois closure of K equals either the symmetric group S_n ($n \neq 4$) or the alternating group A_n

($n \neq 3,5$), then K is a Pólya field if and only if its class-number equals 1, i.e. its ring of integers is a unique factorization domain.

The more general problem of characterizing domains having a regular basis is still undecided (**PROBLEM IV**). Some progress has been made in case of algebraic functions fields K in one variable over a finite field $\mathbf{F}_q$, considered by F.J.VAN DER LINDEN [88]. He noted that the proof of Theorem 5.1 works also for the ring A_P of all elements of K with poles outside P, an arbitrary non-empty set of the set of prime divisors, and used this observation to solve the above problem for all A_P's in case when K is either of genus 0 or an elliptic field. Certain hyper-elliptic fields have been also dealt with. His result is particularly simple for fields K of genus zero. In this case A_P has a regular basis if and only if either it is a unique factorization domain, or the characteristics of K is odd and the class-number of A_P equals 2.

Exercises

1. (H.ZANTEMA [82])   Prove that a quadratic field of discriminant d is a Pólya field if and only if

$$d = 4, 8, -8, -p, q, 4p, 8p, pr,$$

where p, q, r are primes, $p \equiv r \equiv 3 \pmod 4$ and $q \equiv 1 \pmod 4$.

2. (H.ZANTEMA [82])   Show that the maximal real subfield of a cyclotomic field is a Pólya field.

3. Let $R = \mathbf{Z}[i]$ be the ring of integers in the field $\mathbf{Q}(i)$. Show that every polynomial $f \in Int(R)$ of degree ≤ 2 can be expressed as a linear combination with coefficients in R of the polynomials 1, X, and $(X^2 - X)/(1 + i)$.

4. Determine all polynomials in $Int(R)$ of degree not exceeding 8 in the case when R is the ring of integers in $\mathbf{Q}(i)$, $\mathbf{Q}(\sqrt{-3})$ and $\mathbf{Q}(\sqrt{2})$.

VI. Integral-valued derivatives

1. For any positive integer k denote by S_k the set of all polynomials P with rational coefficients such that $P, P', \dots, P^{(k)}$ map $\mathbf{Z}$ into $\mathbf{Z}$, with $P^{(i)}$ being the i-th derivative of P. Let S_∞ be the intersection of all sets S_k. Observe that S_∞ properly contains $\mathbf{Z}[X]$ since the polynomial $\frac{1}{2}(X^4 - X^2)$ belongs to S_∞. Note also that as observed by D.BRIZOLIS [76] the rings $Int(\mathbf{Z}), S_1, S_2, \dots$ are all distinct. In fact one sees easily that the polynomial $\frac{1}{2}(X^2 - X)$ lies in $Int(\mathbf{Z})$ but not in S_1 and a short calculation establishes that for any $n > 1$ if p is a prime exceeding n then the polynomial

$$\frac{(X(X-1)\cdots(X-p+1))^n}{p}$$

lies in $S_{k-1} \setminus S_k$. In the same paper various algebraic properties of S_k have been established.

The following analogue of Theorem 2.1 has been obtained by E.G.STRAUS [51]:

THEOREM 6.1. *The $\mathbf{Z}$-module S_∞ is freely generated by the polynomials*

$$g_i(X) = c_i h_i(X) \quad (i = 1, 2, \dots)$$

where h_i is given by (2.1) and

$$c_i = \prod_p p^{[i/p]}$$

the product being extended over all primes p.

PROOF: Observe that if a polynomial

$$f(X) = \frac{1}{r} X^N + \cdots \quad (r \in \mathbf{Z})$$

belongs to S_∞, then Lemma 2.2 implies that r divides $N!$. Let D_N be the maximal integer r such that $1/r$ is the leading coefficient of a polynomial of degree N belonging to S_∞.

LEMMA 6.2. (i) *If a polynomial $f \in S_\infty$ is of degree N then the denominator of its leading term divides D_N,*

(ii) *If $M < N$, then D_M divides D_N,*

(iii) *If for every* $N \geq 1$ *we choose a polynomial* $f_N \in S_\infty$ *of degree* N *whose leading term equals* $1/D_N$, *then the set* $\{1, f_1, f_2, \ldots\}$ *generates freely the additive group of* S_∞.

PROOF: (i) Let

$$f(X) = \frac{a}{b} X^N + \cdots$$

be a polynomial of degree N belonging to S_∞, with a, b being relatively prime rational integers, b positive. If $m = D_N b/(D_N, b)$, $d_1 = m/D_N$, $d_2 = m/b$, then $(d_1, d_2) = (d_1, a) = 1$ and thus with suitable integers x, y one has $x d_1 + y d_2 a = 1$. It follows that the polynomial $x f_N + y f$ is of degree N, lies in S_∞ and its leading coefficient equals $1/m$, so we get $m \leq D_N \mid m$, and hence $m = D_N$, which implies $b \mid D_N$.

(ii) The polynomial $X^{N-M} f_M(X)$ is of degree N, belongs to S_∞ and has $1/D_M$ for its leading coefficient. The assertion results now from (i).

(iii) One applies induction based on the observation that if $f \in S_\infty$ is of degree N and has a/b $(a, b \in \mathbf{Z})$ for its leading coefficient, then (i) shows that $c = a D_N/b$ is an integer and thus $f - c f_N$ lies in S_∞. As its degree is smaller than N we may apply the inductional assumption to get

$$f = c f_N + \sum_{j \leq N-1} \alpha_j f_j$$

with suitable $\alpha_j \in \mathbf{Z}$. $\square$

The next lemma provides an upper bound for prime power divisors of D_N:

LEMMA 6.3. *Let* $N \geq 1$. *If* p *is a prime,* $k \geq 1$ *and* p^k *divides* D_N, *then* $p^k \mid [N/p]!$, $p \leq \sqrt{N}$, *and*

$$k \leq \sum_{i \geq 2} \left[\frac{N}{p^i} \right].$$

PROOF: Let M be the smallest positive integer for which $f = M f_N$ has integral coefficients. Clearly M is a multiple of D_N, and our choice of M implies that not all coefficients of f are multiples of p. Thus $F = f \bmod p$ is not the zero polynomial and it follows that for any integer i the multiplicity of i as a root of F is well-defined. Denote this multiplicity by $m(i)$. If we write

(6.1) $$f(X) = (X - i)^{m(i)} g(X) + p h(X),$$

with $g, h \in \mathbf{Z}[X]$ and $g(i)$ not divisible by p then it follows that with a suitable $A \in \mathbf{Z}[X]$ one has

$$f^{(m(i))}(X) = m(i)! g(X) + (X - i) A(X) + p h^{(m(i))}(X),$$

The product of m consecutive integers being always divisible by $m!$, we obtain with a suitable $h_1 \in \mathbf{Z}[X]$

$$h^{(m(i))}(X) = m(i)! h_1(X),$$

and thus

$$0 \equiv f^{(m(i))}(i) \equiv m(i)!(g(i) + ph_1(i)) \pmod{p^k},$$

which implies the divisiblity of $m(i)!$ by p^k. Denoting by m the minimal value of $m(i)$ we get $p^k \mid m!$ and (6.1) implies the divisibility of F by $(X - i)^m$ for $i = 0, 1, \ldots, p - 1$. This gives

$$N = \deg f \geq \deg F \geq pm$$

and thus $m \leq [N/p]$ leading to $p^k \mid m! \mid [N/p]!$. Now $p \leq \sqrt{N}$ follows immediately and the last assertion results from the fact that the exponent of the maximal power of a prime p dividing $n!$ equals $\sum_{i \geq 1} [n/p^i]$. $\square$

COROLLARY. *For $N \geq 1$ one has*

$$D_N \leq \frac{N!}{c_N}.$$

PROOF: The lemma implies

$$D_N \leq \prod_p p^{\sum_{i \geq 2} [N/p^i]} = \frac{N!}{c_N}. \quad \square$$

The above corollary shows that the leading coefficient of g_N is not smaller than D_N. To conclude the proof of the theorem it is thus sufficient to check that all polynomials g_i lie in S_∞.

Theorem 2.1 implies that each polynomial g_i maps $\mathbf{Z}$ into $\mathbf{Z}$ and we shall now prove that the same is true for all their derivatives. Clearly this holds for $g_0 = 1$. Assume that the polynomials $g_0, g_1, \ldots, g_{N-1}$ lie in S_∞.

Put $A_{-1} = 1$ and

$$A_n(X) = \prod_{j=0}^{n} (X - j) \quad (n \geq 0).$$

We shall prove the following elementary identity:

$$(6.2) \qquad \frac{dA_{n-1}(X)}{dX} = \sum_{k=0}^{n-1} (-1)^k \frac{n(n-1)\ldots(n-k)}{1+k} A_{n-k-2}(X) \quad (n \geq 1).$$

Since it is obviously true for $n = 1$ assume that it holds for all integers less than n. Then we have

$$A'_{n-1}(X) - A_{n-2}(X) = A'_{n-2}(X)(X - n + 1)$$

$$= \sum_{k=0}^{n-2} (-1)^k \frac{(n-1)(n-2)\cdots(n-k-1)}{k+1} A_{n-k-3}(X)(X - n + k + 2 - (k+1))$$

$$= \sum_{k=0}^{n-2} (-1)^k \frac{(n-1)(n-2)\cdots(n-k-1)}{k+1} A_{n-k-2}(X)$$

$$+ \sum_{k=0}^{n-2} (-1)^{k+1} (n-1)(n-2)\cdots(n-k-1) A_{n-k-3}(X)$$

$$= \sum_{k=0}^{n-2} (-1)^k \frac{(n-1)(n-2)\cdots(n-k-1)}{k+1} A_{n-k-2}(X)$$

$$+ \sum_{k=1}^{n-1} (-1)^k (n-1)(n-2)\cdots(n-k-1) A_{n-k-2}(X) = (n-1) A_{n-2}(X)$$

$$+ \sum_{k=1}^{n-2} (-1)^k \frac{n(n-1)(n-2)\cdots(n-k)}{k+1} A_{n-k-2}(X) + (-1)^n (n-1)!$$

and (6.2) follows immediately.

Since $h_j(X) = A_{j-1}(X)/j!$, the formula (6.2) leads to

$$g'_n(X) = c_n h'_n(X) = \frac{c_n}{n!} \frac{dA_{n-1}(X)}{dX}$$

$$= c_n \sum_{k=1}^n \frac{(-1)^k}{k} h_{n-k}(X) = \sum_{k=1}^n \frac{(-1)^k}{k} \frac{c_n}{c_{n-k}} g_{n-k}(X)$$

and it remains to check that the numbers

$$\frac{c_n}{kc_{n-k}} = \frac{\prod_{p \le n} p^{[n/p]-[(n-k)/p]}}{k}$$

are integers for $k = 1, 2, \ldots, n$. If we write

$$\frac{c_n}{kc_{n-k}} = \prod p^{\alpha_p},$$

and p does not divide k then evidently $\alpha_p \ge 0$ and if p^β is the highest power of p dividing k and $\beta > 1$, then we get

$$\alpha_p = [n/k] - [(n-k)/p] - \beta = \frac{k}{p} - \beta \ge p^{\beta-1} - \beta \ge 0. \quad \square$$

2. It turns out that the set of polynomials occuring in Theorem 6.1 will not change if we replace here the derivatives by consecutive differences. More precisely, L.CARLITZ [59] obtained the following result:

If $M_0 = Int(\mathbf{Z})$ and for $k = 1, 2, \ldots$ we define M_k to be the set of all polynomials $f \in \mathbf{Q}[X]$ having the property that for all integers n the polynomial

$$\frac{f(X + n) - f(X)}{n}$$

belongs to M_{k-1}, then the intersection M of all sets M_k coincides with S_∞.

This result has been rediscovered by V.LAOHAKOSOL, P.UBOLSRI [80]. The analogue of it for algebraic number fields fails in general, as noted by J.-L.CHABERT [93] who characterized Dedekind domains for which it is true, correcting an assertion of D.BARSKY [72]. Cf. D.BARSKY [73] (where the existence of a regular basis for M has been studied) and Y.HAOUAT, F.GRAZZINI [77], [78], [79].

Prime ideals of $M = S_\infty$ have been studied by P.J.CAHEN [75] and a description of maximal ideals in the rings S_k and S_∞ has been given by D.BRIZOLIS [76] (Theorems 1.1.7 and 1.2.2). The analogue of the ring M in the case of polynomial rings has been dealt with by C.G.WAGNER [76] and Y.HAOUAT [86].

3. If in the definition of the ring M_1 one does not assume that f is a polynomial then one arrives at a class of functions considered by N.G.DE BRUIJN [55] and called by him *modular functions*. An integer-valued function f is thus modular, provided for all $x \in \mathbf{Z}$ and $n = 1, 2, \ldots$ the difference $f(x + n) - f(x)$ is divisible by n. De Bruijn showed that a function is modular if and only if it can be written in the form

$$c_0 + \sum_{k \geq 1} c_k F_k,$$

with rational integral c_k's, where

$$F_k(X) = s_k \frac{(X + r)(X + r - 1) \cdots (X + r - k + 1)}{k!},$$

$r = [k/2]$ and s_k denotes the least common multiple of the first k positive integers.

If in the definition of modular functions one restricts the defining property to positive values of x, then one gets another class of functions, called *universal functions* in N.G.DE BRUIJN [55] or *pseudo-polynomials* in R.R.HALL [71]. An example of such a function which is not a polynomial is given by $f(n) = [en!]$. De Bruijn showed that every such function can be written in the form

$$c_0 + \sum_{k \geq 1} c_k s_k \binom{X}{k},$$

(with $c_k \in \mathbf{Z}$ and s_k as above) and Hall after giving the same description proved that the set of all pseudo-polynomials forms an integral domain, which is not a unique factorization domain. He showed also that any pseudo-polynomial f satisfying

$$f(n) = O(a^n)$$

with a certain $a < e - 1$ is necessarily a polynomial. Cf. also U.RAUSCH [87] and I.RUZSA [71]. It has been shown by M.V.SUBBARAO [66] that a universal function f which is multiplicative (i.e. satisfies $f(mn) = f(m)f(n)$ for relatively prime m, n) equals either 0 or n^k with a suitable k. Another proof can be found in A.SOMAYAJULU [68]. The assumptions in this theorem have been later essentially weakened by A.IVÁNYI [72], B.M.PHONG, J.FÉHER [90] and B.M.PHONG [91].

A more general class of functions appears in E.G.STRAUS [52].

The analogue for arbitrary rings R has been considered by W.NÖBAUER [76], who called a function $f : R \longrightarrow R$ compatible if for all ideals I of R the congruence $a \equiv b \pmod{I}$ implies $f(a) \equiv f(b) \pmod{I}$. He determined a.o. the structure of the semigroup (with regard to composition) of all compatible function in $R = \mathbf{Z}/n\mathbf{Z}$. This notion has been also regarded in greater generality in the theory of universal algebras. See e.g. A.L.FOSTER [67],[70].

An analogue of Theorem 6.1 for rings of integers in algebraic number fields has been proved by K.ROGERS, E.G.STRAUS [85]. They obtained the following result:

For every $j = 0, 1, 2, \ldots$ there is an ideal A_j of $\mathbf{Z}_K$ containing $j!$ and a sequence $b_0, b_1, \ldots$ of elements of $\mathbf{Z}_K$ such that if we put

$$f_j(X) = -b_0(X - b_1) \cdots (X - b_j),$$

then an n-th degree polynomial f and all its derivatives map $\mathbf{Z}_K$ into $\mathbf{Z}_K$ if and only if with suitable $a_j \in A_j$ $(j = 0, 1, 2, \ldots, n)$ one has

$$n!f(X) = \sum_{j=0}^{n} a_j f_j(X).$$

The same question has been considered for discrete valuation rings by P.J.CAHEN, J.L.CHABERT [71] and for arbitrary Noetherian domains by J.L. CHABERT [79b].

The sets S_k and their analogues for other rings turned out to be more complicated than S_∞. A description of S_1 has been given by D.BRIZOLIS and E.G.STRAUS [76]. They showed that S_1 is a free $\mathbf{Z}$-module and described a basis for it, whose first elements are $1, X, 2\binom{X}{2}, 6\binom{X}{3}, 6\binom{X}{4} + \binom{X}{2}, \ldots$.

No description of S_k in case $k \geq 2$ is known (**PROBLEM V**). The same question may be asked also for rings other than $\mathbf{Z}$ (**PROBLEM VI**).

Exercises

1. Prove that the function $f(n) = [en!]$ is a pseudo-polynomial.

2. (E.G.STRAUS [51]) Let m be a fixed integer. Show that every polynomial $f \in \mathbf{Q}[X]$ which at points $0, 1, \ldots, m - 1$ assumes integral values with all its derivatives can be written as a linear combination with integral

coefficients of the polynomials

$$f_{mn}(X)$$
$$= \frac{(X(X-1)\cdots(X-m+1))^{[n/m]}X(X-1)\cdots(X-n+m[n/m]+1)}{D_{mn}},$$

where

$$D_{mn} = \prod_{p<m} p^{[n/p]!p} \prod_{p\geq m} p^{[n/m]!p}.$$

3. (D.BRIZOLIS, E.G.STRAUS [76]) Let p be a prime, let $\mathbf{Q}_{(p)}$ be the ring of all rational numbers with denominators not divisible by p and let A_p be the set of all $f \in \mathbf{Q}[X]$ satisfying $f(\mathbf{Q}_{(p)}) = \mathbf{Q}_{(p)}$ and $f'(\mathbf{Q}_{(p)}) = \mathbf{Q}_{(p)}$.

(i) Show that A_p has a regular basis, i.e. it is a free $\mathbf{Q}_{(p)}$-module which has a basis f_n $(n = 0, 1, 2, \dots)$ satisfying $\deg f_n = n$.

(ii) Prove that the subring of A_2 consisting of polynomials having their degrees ≤ 6 is a free $\mathbf{Q}_{(p)}$-module generated by

$$1, \quad X, \quad 2\binom{X}{k} + \binom{X}{k-2} \quad (k = 2, 3, 4, 5), \quad 4\binom{X}{6} + 2\binom{X}{2}.$$

(iii) Obtain the analogue of (ii) in the cases $p = 3$ and $p = 5$.

4. Prove that every polynomial $f \in Int(\mathbf{Z})$ of degree ≤ 5 satisfying $f' \in Int(\mathbf{Z})$ can be uniquely expressed as a linear combination with integral coefficients of the polynomials

$$1, \quad X, \quad X(X-1), \quad X(X-1)(X-2), \quad 6\binom{X}{4} + \binom{X}{2}, \quad 30\binom{X}{5} + 3\binom{X}{3}.$$

VII. Algebraic properties of Int(R)

1. Algebraic properties of the rings $Int(R)$ have been intensively studied, mostly in the case $R = \mathbf{Z}$. We prove now a sample of relevant results and start with two negative statements.

THEOREM 7.1. (i) *The ring $Int(\mathbf{Z})$ is not Noetherian.*

(ii) *The ring $Int(\mathbf{Z})$ is not a Bezout ring, i.e. not every its finitely generated ideal is principal.*

PROOF: (i) Let $p_1 < p_2 < p_3 < \ldots$ be the sequence of all primes and for $k = 1, 2, \ldots$ denote by A_k the ideal of $Int(\mathbf{Z})$ generated by $h_1, h_2, \ldots, h_{p_k}$, where h_i is given by (2.1). Clearly the A_k's form an ascending sequence and it suffices to show that they are all distinct. If for some k one has $A_k = A_{k-1}$, then with suitable polynomials $V_1, \ldots, V_k$ from $Int(\mathbf{Z})$ we get

$$\frac{X(X-1)\cdots(X - p_k + 1)}{p_k!} = \sum_{i=1}^{k-1} \frac{X(X-1)\cdots(X - p_i + 1)}{p_i!} V_i(X).$$

Dividing by X and putting $X = p_k$ we obtain on the right hand-side a rational number whose denominator divides $p_{k-1}!$ and thus is prime to p_k, whereas the left hand-side equals $\dfrac{1}{p_k}$. This is a clear contradiction.

(ii) The assertion follows by observing that the ideal generated in $Int(\mathbf{Z})$ by 2 and X is not principal. $\square$

The assertion (i) is a particular case of the following result of R.GILMER, W.HEINZER, D.LANTZ [92]:

If R is a Noetherian domain which is either one-dimensional or integrally closed then the ring $Int(R)$ is Noetherian if and only if $Int(R) = R[X]$.

In the same paper an example of a domain R with Noetherian $Int(R) \neq R[X]$ has been given: $R = k + M^2$, where k is a finite field and M is the maximal ideal of the ring of power series in two variables over k.

A characterization of rings R for which $Int(R)$ is Noetherian is not known **(PROBLEM VII)**.

2. Let R be a domain and let A be a subring of $Int(R)$ containing $R[X]$. The ring A is called a *Skolem ring* (associated with R) if for every finite sequence

$f_1, \ldots, f_n$ of polynomials belonging to A which for every $r \in R$ satisfies

$$\sum_{i=1}^{n} f_i(r)R = R$$

one has

$$\sum_{i=1}^{n} f_i A = A.$$

This condition (the *Skolem property*) can be also stated in the following equivalent way:

If I is a finitely generated ideal of A such that for every $a \in R$ there is at least one polynomial $f \in I$ with $f(a) = 1$, then $I = A$.

The Skolem property has been first considered by T.SKOLEM ([36], [37a]) who showed that $Int(\mathbf{Z})$ and its analogue for polynomials of several variables possess it. The general case occurs first in D.BRIZOLIS [75],[76]. Certain related properties have been considered by D.BRIZOLIS [79], J.-L.CHABERT [83],[88] and D.L.MCQUILLAN [85a].

3. We prove now a generalization of Skolem's result to algebraic number fields, due to D.BRIZOLIS [76]:

THEOREM 7.2. *If K is an algebraic number field and R is its ring of integers, then $Int(R)$ is a Skolem ring.*

PROOF: Let $f_1, \ldots, f_n \in Int(R)$ and assume that for every $a \in R$ we have

$$\sum_{i=1}^{n} f_i(a)R = R.$$

Observe first that the polynomials f_i cannot have a common zero in any extension of K. Indeed, if u were such a zero then u would be an an algebraic number, and its minimal (over K) polynomial g would divide every polynomial f_i. Getting rid of the denominators we would get

$$cf_i = gh_i \quad (i = 1, 2, \ldots, n)$$

with suitable $c \in R$ and $h_i \in Int(R)$. For every $a \in R$ the number $g(a)$ divides $cf_i(a)$ for $i = 1, 2, \ldots, n$, thus

$$cR = c\sum_{i=1}^{n} f_i(a)R \subset g(a)R,$$

so $g(a)$ divides c and this gives a contradiction, since the polynomial g is non-constant. It follows that with suitable polynomials $g_1, \ldots, g_n \in R[X]$ and $b \in R$ we have

(7.1) $$\sum_{i=1}^{n} f_i(X)g_i(X) = b.$$

If b is a unit, then our assertion becomes evident, otherwise we can write

$$bR = \prod_{j=1}^{r} P_j^{e_j},$$

where P_j are prime ideals of R and the e_j's are positive integers. Let N_j be the absolute norm of the ideal P_j, i.e. $N_j = \#R/P_j$, put $A_{j1} = 1$,

$$A_{js} = \prod_{k=1}^{s-1} (f_s^{N_j-1} - f_k^{N_j-1}) \quad (j = 1, 2, \ldots, r; \; s = 2, 3, \ldots, n),$$

and

$$A_j = \sum_{i=1}^{n} A_{ji} f_i, \quad (j = 1, 2, \ldots, r).$$

The property of A_j that we need is stated in the next lemma:

LEMMA 7.3. *For every $a \in R$ we have $A_j(a) \notin P_j$.*

PROOF: Note that if $u \in R$, then by the analogue of Fermat's theorem in algebraic number fields $u^{N_j-1} \bmod P_j$ equals 0 if $u \in P_j$ and equals 1 otherwise. This implies that $A_{js}(a) \bmod P_j$ is non-zero only in the following two cases:

$$a) \quad f_1(a), \ldots, f_{s-1}(a) \in P_j, \; f_s(a) \notin P_j,$$

in which case we have $A_{js}(a) \bmod P_j = 1$, and

$$b) \quad f_1(a), \ldots, f_{s-1}(a) \notin P_j, f_s(a) \in P_j,$$

in which case $A_{js}(a) \bmod P_j = (-1)^{s-1}$.

We see hence that $A_{js}(a) f_s(a) \bmod P_j$ is non-zero only if s is the smallest index for which $f_s(a) \notin P_j$ and the assertion follows. $\square$

Now let $\pi_j \in P_j$, $\pi_j \notin P_i$ (for $j \neq i$), $\pi = \pi_1 \cdots \pi_r$ and

$$g(X) = \sum_{j=1}^{r} \frac{\pi}{\pi_j} A_j(X).$$

The lemma implies that for every $a \in R$ the ideals $g(a)R$ and bR are relatively prime, hence $g(a)$ is invertible in the factor-ring R/bR. If thus T denotes the cardinality of the multiplicative group of all invertible elements of R/bR then we have $g(a)^T \equiv 1 \pmod{bR}$ for all $a \in R$. This remark shows that the polynomial

$$g_0(X) = \frac{g(X)^T - 1}{b}$$

lies in $Int(R)$. Finally put

$$F_i = \left(\sum_{j=1}^{r} \frac{\pi}{\pi_j} A_{ji} \right) g^{T-1} - g_0 g_i,$$

and use (7.1) to obtain

$$\sum_{i=1}^{n} F_i f_i = g^T - b g_0 = 1.$$

Since all polynomials F_j belong to $Int(R)$, this equality shows that the ideal generated by the f_i's coincides with $Int(R)$. $\square$

COROLLARY. (T.SKOLEM [36]) *The ring $Int(\mathbf{Z})$ is a Skolem ring.* $\square$

4. A description of domains R which have an associated Skolem ring does not seem to be known in the general case (**PROBLEM VIII**). For Dedekind domains such a description has been given by D.BRIZOLIS [75], for domains with the finite norm property by D.L.McQUILLAN ([78], Theorem 2) and for Noetherian domains by J.-L.CHABERT [78]. These descriptions connect Skolem rings with *D-rings*, defined as domains in which every non-constant polynomial attains at least one non-invertible value (H.GUNJI, D.L.McQUILLAN [75] . Such rings have been called *rings with property S* by D.BRIZOLIS [75]). We shall deal with them in the next section.

D.BRIZOLIS [75] showed that a Dedekind domain R has an associated Skolem ring if and only if R is a D-ring and for every maximal ideal M of R the quotient field R/M is either finite or algebraically closed. A result of D.L.McQUILLAN [78] deals with domains R having the finite norm property and states that such a domain has an associated Skolem ring if and only if R is either a D-ring or an algebraically closed field. Later J.-L.CHABERT [78] showed that if R is Noetherian that it has an associated Skolem ring if and only if the largest possible ring, which can be a Skolem ring, associated with R, i.e. the ring $Int(R)$ is in fact Skolem and proved that this happens if and only if R is a D-ring, every its prime ideal is an intersection of maximal ideals (i.e. R is a *Jacobson ring* or *Hilbert ring* (see O.GOLDMAN [51], W.KRULL [51])) and finally for every maximal ideal M of R either R/M is finite or M is of height one and R/M is algebraically closed.

Examples of non-Noetherian rings R for which $Int(R)$ is a Skolem ring have been given by J.-L.CHABERT [79a]. One of them is $R = Int(\mathbf{Z})$.

The smallest ring which could be a Skolem ring associated with a given domain R is $R[X]$. Already T.SKOLEM [36] noted that $\mathbf{Z}[X]$ is not a Skolem ring. In fact, for every integer n the numbers 3 and $n^2 + 1$ are relatively prime, however there cannot exist polynomials $A, B \in \mathbf{Z}[X]$ such that

$$3A(X) + (X^2 + 1)B(X) = 1,$$

since otherwise this equality would hold also for complex arguments and putting in it $X = i$ we would obtain that 3 is invertible in the ring $\mathbf{Z}[i]$.

It has been established by D.L.McQUILLAN ([78], Theorem 1) that if R has the finite norm property then $R[X]$ is a Skolem ring if and only if R is an algebraically closed field, and J.-L.CHABERT [79a] characterized domains R for which $R[X]$ is a Skolem ring as such which are Jacobson rings and in which every quotient field R/M ($M \subset R$ - maximal ideal) is algebraically closed.

5. The property defining Skolem rings can be also considered in a much more general setting: let R be a domain, let A be a given set and let D be a ring of maps $A \to R$. For any ideal I of D and any $a \in R$ put $I(a) = \{f(a) : f \in I\}$. One says that D has the *Skolem property*, provided for every finitely generated proper ideal I of D there exists $a \in A$ such that $I(a) \notin R$. In case $D = Int(R)$ we regain the definition of the Skolem ring.

Another interesting case arises when D is the ring of all R-valued polynomials in k variables with coefficients in the field of fractions of R and $A = R^k$. For $R = \mathbf{Z}$ this has been considered already by T.SKOLEM [36] who established the Skolem property in this case and D.BRIZOLIS [75] described Dedekind domains R for which this happens: R has to be a D-ring and for every maximal ideal M of R the field R/M is either finite or algebraically closed. In particular the ring of all integers in an algebraic number field has this property. It has been later shown by J.-L.CHABERT [83] that if R is a Noetherian domain then the ring D of R-valued polynomials in k variables has the Skolem property if and only if the following analogue of Hilbert's Nullstellensatz (Hilbert's theorem on zeros) holds in D:

If I is a finitely generated ideal of D, $f \in D$ and for every $a = [r_1, \ldots, r_k] \in R^k$ one has $f(r_1, \ldots, r_k) \in I(a)$, then some power of f lies in I.

He gave also two other properties equivalent to the Skolem property in this case.

Nothing seems to be known about the Skolem property for other sets A **(PROBLEM IX)**.

6. The next theorem describes prime and maximal ideals of $Int(\mathbf{Z})$.

THEOREM 7.4. (D.BRIZOLIS [76]). (i) *The maximal ideals of $Int(\mathbf{Z})$ are in a one-to-one correspondence with pairs $[p, c]$, where p is a rational prime and c is a p-adic integer. This correspondence is given by*

$$[p, c] \Longleftrightarrow M(p, c) = \{f : f \in Int(\mathbf{Z}), f(c) \in p\mathbf{Z}_p\}.$$

(ii) *Every non-maximal prime ideal of $Int(\mathbf{Z})$ is of the form*

$$P_g = g(X)\mathbf{Q}[X] \cap Int(\mathbf{Z}),$$

where $g \in \mathbf{Z}[X]$ is an irreducible polynomial. Two ideals P_g and P_h coincide if and only if the polynomials g, h differ by a constant factor.

PROOF: A description of prime ideals of $\mathbf{Z}[X]$ (due to L.KRONECKER [KR]) is needed first:

LEMMA 7.5. *Let P be a non-zero prime ideal of $\mathbf{Z}[X]$, $R = \mathbf{Z}[X]/P$ and let $p \geq 0$ be the characteristics of the field of quotients of R. Then we have*

$$P = p\mathbf{Z}[X] + f\mathbf{Z}[X].$$

Here f is in the case $p = 0$ a polynomial irreducible over $\mathbf{Q}$ and in the case $p > 0$ either zero or a polynomial whose reduction mod p is irreducible over $\mathbf{F}_p$.

Moreover $p > 0$ holds if and only if $P \cap \mathbf{Z} \neq 0$ and P is maximal if and only if $p > 0$ and $f \neq 0$.

PROOF: Let $T : \mathbf{Z}[X] \rightarrow R = \mathbf{Z}[X]/P$ be the canonical map and $t : \mathbf{Z} \rightarrow R$ its restriction to $\mathbf{Z}$. If $T(X) = a$, then for every $f \in \mathbf{Z}[X]$ we have $T(f) = \hat{f}(a)$, where for $f = \sum_{k=0}^{N} a_k X^k$ we define

$$\hat{f} = \sum_{k=0}^{N} t(a_k) X^k,$$

and thus

$$P = \{f : \hat{f}(a) = 0\}.$$

Let k be the field of quotients of R and let $p \geq 0$ be its characteristics. If $p = 0$, then a is an algebraic number, t is an injection (hence $P \cap \mathbf{Z} = 0$) and we see that P consists of all multiples of the minimal polynomial of a. If $p \neq 0$, then $p \in P$, t maps $\mathbf{Z}$ onto $\mathbf{F}_p$ and either a lies in a finite extension of $\mathbf{F}_p$ in which case P consists of all polynomials over $\mathbf{Z}$ whose reductions mod p are divisible by the minimal polynomial of a over $\mathbf{F}_p$ or a is transcendental over $\mathbf{F}_p$, in which case P consists of all polynomials having their coefficients divisible by p. The remaining assertions follow now immediately. $\square$

(A constructive description of all ideals of $\mathbf{Z}[X]$ has been given in G.SZE-KERES [52]. See also F.CHATELET [67]. A similar result for the ring $\mathbf{Z}[X, Y]$ gave P.G.TROTTER [78]).

We return to the proof of the theorem. Obviously $M(p, c)$ is an ideal of $Int(\mathbf{Z})$. To obtain its maximality observe first that since $\mathbf{Z}$ is dense in $\mathbf{Z}_p$, the ring of p-adic integers, every polynomial $f \in Int(\mathbf{Z})$ maps $\mathbf{Z}_p$ into $\mathbf{Z}_p$. This observation shows that the map

$$T : Int(\mathbf{Z}) \rightarrow \mathbf{Z}/p\mathbf{Z}$$

given by

$$T(f) = f(c) \bmod p$$

is well-defined, surjective and its kernel equals $M(p, c)$, thus $M(p, c)$ is a maximal ideal.

Now let M be an arbitrary maximal ideal of $Int(\mathbf{Z})$. We prove first that $M \cap \mathbf{Z}$ is non-zero. Indeed, otherwise the prime ideal $M \cap \mathbf{Z}[X]$ of $\mathbf{Z}[X]$ would be equal to $f\mathbf{Z}[X]$ for a certain polynomial f over $\mathbf{Z}$, irreducible over $\mathbf{Q}$. If $g \in M$, then for an integer N we would have $Ng \in M \cap \mathbf{Z}[X] = f\mathbf{Z}[X]$, and thus the quotient g/f would be a polynomial over $\mathbf{Q}$. This leads to

$$M \subset f\mathbf{Q}[X] \cap Int(\mathbf{Z}),$$

and since $f\mathbf{Q}[X] \cap Int(\mathbf{Z})$ is a proper ideal of $Int(\mathbf{Z})$, the maximality of M gives

$$M = f\mathbf{Q}[X] \cap Int(\mathbf{Z}).$$

Let p be a prime which does not divide the leading coefficient nor the discriminant of f and for which the congruence

$$f(x) \equiv 0 \pmod{p}$$

is solvable. Hensel's lemma shows that f has a root c in $\mathbf{Z}_p$, hence $f \in M(p,c)$ and $M \subset M(p,c)$ follows. But p does not lie in M and finally we obtain that M is contained properly in $M(p,c)$, contradicting its maximality.

Thus $M \cap \mathbf{Z} \neq 0$. Denote by p the minimal positive integer in $M \cap \mathbf{Z}$. Clearly p is a prime. Since $Int(\mathbf{Z})$ is countable, so is M, and we can enumerate all non-constant elements of M: $f_1, f_2, \ldots$. Let M_n be for $n = 1, 2, \ldots$ the ideal generated by $p, f_1, \ldots, f_n$. Since $M_n \neq Int(\mathbf{Z})$ and $Int(\mathbf{Z})$ is a Skolem ring by the Corollary to Theorem 7.2, there is a rational integer a_n such that all numbers $f_1(a_n), \ldots, f_n(a_n), p$ have a common divisor > 1, thus $f_i(a_n)$ is divisible by p for $i = 1, 2, \ldots, n$. The sequence $\{a_n\}$ has a limit point, say c in $\mathbf{Z}_p$ and by continuity we arrive at

$$f_i(c) \in p\mathbf{Z}_p$$

fpr $i = 1, 2, \ldots$, thus $M = M(p,c)$, as asserted.

To obtain (i) it remains to show that distinct pairs $[p,c]$ lead to distinct prime ideals. This will follow immediately from the following lemma:

LEMMA 7.6. *If a, b are distinct p-adic integers, then there exists a polynomial $f \in Int(\mathbf{Z})$ satisfying*

$$f(a) \in p\mathbf{Z}_p, \quad f(b) \notin p\mathbf{Z}_p.$$

PROOF: Let $q = p^m$ be the highest power of p dividing $a - b$ and consider the polynomial

$$h(X) = \frac{(X - a)(X - a - 1) \cdots (X - a - q + 1)}{q!},$$

which maps $\mathbf{Z}_p$ into $\mathbf{Z}_p$.

Obviously $h(a) = 0$ and to compute $h(b)$ write $b = a + uq$ with a p-adic unit u and observe that all terms of the product

$$h(b) = \prod_{j=0}^{q-1} \frac{uq - j}{q - j}$$

are p-adic units and thus $h(b)$ is not divisible by p. If now N is a sufficiently large integer and $\alpha \in \mathbf{Z}$ satisfies $\alpha \equiv a \pmod{p^N}$ then the polynomial

$$f(X) = \frac{(X - \alpha)(X - \alpha - 1) \cdots (X - \alpha - q + 1)}{q!}$$

satisfies our needs. $\square$

Now we turn to (ii). As P_g is the intersection of a maximal ideal of $\mathbf{Q}[X]$ with $Int(\mathbf{Z})$, it is a prime ideal and (i) shows that it cannot be maximal. Let P be a prime non-maximal ideal of $Int(\mathbf{Z})$. If its intersection with $\mathbf{Z}$ is non-zero, then the argument used in (i) leads to $P = M(p,c)$ with suitable p and c, hence P is maximal, contrary to our assumption. Thus $P \cap \mathbf{Z} = 0$. There is a polynomial $g \in \mathbf{Z}[X]$ such that the ideal $I = P\mathbf{Q}[X]$ of $\mathbf{Q}[X]$ equals $g\mathbf{Q}[X]$. Now it suffices to show that I is a prime ideal of $\mathbf{Q}[X]$ and $P = I \cap S[\mathbf{Z}]$. First we show that I is a proper ideal. Were it not so, then we could find $f_1, \ldots, f_r \in P$ and $g_1, \ldots, g_r \in \mathbf{Q}[X]$ with

$$\sum_{j=1}^{r} f_j(X)g_j(X) = 1,$$

and multiplying by the common denominator of the coefficients of the polynomials appearing on the left-hand side of this equality we would obtain a non-zero rational integer lying in P, which is not possible. If I would not be prime, then the polynomial g would be reducible, thus $g = g_1 g_2$ with suitable non-constant polynomials $g_1, g_2 \in \mathbf{Z}[X]$. Since g_1, g_2 do not lie in $I \supset P$ this contradicts the primality of P.

Finally let f be an arbitrary element of $I \cap Int(\mathbf{Z})$. Then with a suitable non-zero $d \in \mathbf{Z}$ we can write

$$df = \sum_{j=1}^{s} \varphi_j \psi_j$$

with $\varphi_j \in P$ and $\psi_j \in \mathbf{Z}[X]$ we get $df \in P$, which in view of $d \notin P$ leads to $f \in P$. Thus $I \cap Int(\mathbf{Z}) \subset P$ and since the reverse inclusion is evident, the first part of the assertion (ii) follows. The second part being obvious we are ready. $\square$

COROLLARY 1. *The ring $Int(\mathbf{Z})$ has uncountably many maximal ideals.*

PROOF: It suffices to note that for every prime p the ring $\mathbf{Z}_p$ is uncountable. $\square$

COROLLARY 2. *The Krull dimension of $Int(\mathbf{Z})$ equals 2.*

(Recall, that the *Krull dimension* of a ring R is defined as the maximal number N for which there exists a chain $P_0 \subset P_1 \subset \ldots \subset P_N$ of distinct prime ideals of R).

PROOF: It suffices to observe that $P_g \subset P_h$ implies $P_g = P_h$. $\square$

A more general result has been established by J.-L.CHABERT [77], who showed that if R is a Noetherian domain, then $\dim Int(R) = 1 + \dim R$. For an arbitrary domain R the inequality $\dim Int(R) \geq \dim R[X] - 1$ has been established by P.-J.CAHEN [90]. He gave examples with equality holding here and showed moreover that for *Jaffard domains*, i.e domains R with $\dim R[X_1, \ldots, X_n] = \dim R + n$ for $n = 1, 2, \ldots$ (see D.F.ANDERSON, A.BOUVIER, D.E.DOBBS, M.FONTANA, S.KABBAJ [88]) one has $\dim Int(R) = \dim R[X]$.

7. The analogue of Theorem 7.4 holds also for all Dedekind domains R with finite norm property for which $Int(R)$ is a Skolem ring (D.BRIZOLIS [76], Theorem 2.2.3). A similar description of prime ideals of $Int(R)$ for arbitrary

Noetherian domains has been given by J.-L.CHABERT [77],[78] (For a particular case see J.-L.CHABERT [71], P.-J.CAHEN [91]). The case of a pseudovaluation domain has been treated in P.-J.CAHEN, Y.HAOUAT [88]. Prime and maximal ideals of $Int(A, R)$ for R-fractional subsets A of the quotient field of R (i.e. sets for which with a suitable non-zero $a \in R$ one has $aA \subset R$) have been described by D.L.McQUILLAN [85b],[85c] in the case of a Dedekind domain R.

8. A domain R with the property that for every maximal ideal M of R the localization R_M is a *valuation domain*, i.e. for any non-zero elements a, b of R_M either a/b or b/a lies in R_M, is called a *Prüfer domain*. Recall that the *localization R_M* is defined by

$$R_M = \{a/b : a, b \in R, b \notin M\}.$$

One can define Prüfer domains equivalently by demanding the invertibility of all non-zero finitely generated ideals of R (see e.g. [MIT], Theorem 22.1). In particular Noetherian Prüfer domains coincide with Dedekind domains.

It has been shown by D.BRIZOLIS [79] that $Int(\mathbf{Z})$ is a Prüfer domain. This surprises a little, since, as we shall now see, $\mathbf{Z}[X]$ is not a Prüfer domain.

THEOREM 7.7. $\mathbf{Z}[X]$ *is not a Prüfer domain.*

PROOF: The ring $\mathbf{Z}[X]$ is a Noetherian unique factorization domain hence if it were a Prüfer domain, then it would be a Dedekind domain and thus a principal ideal domain. But the ideal $2\mathbf{Z}[X] + X\mathbf{Z}[X]$ is obviously non-principal. $\square$

The following argument shows directly that for any prime p the ideal $M = p\mathbf{Z}[X] + X\mathbf{Z}[X]$ of $\mathbf{Z}[X]$ is not invertible:

Assuming that M is invertible, we obtain the existence of a finitely generated ideal

$$I = \sum_{j=1}^{n} f_j \mathbf{Z}[X]$$

such that the product MI is principal and equal to $g\mathbf{Z}[X]$, say. Then

$$g\mathbf{Z}[X] = MI = \sum_{j=1}^{n} pf_j \mathbf{Z}[X] + \sum_{j=1}^{n} X f_j \mathbf{Z}[X],$$

thus all polynomials pf_j, Xf_j are divisible by g. We have two cases to consider:

a) $g(0) = 0$, i.e. X divides g. In this case with a suitable h in $\mathbf{Z}[X]$ we have $g(X) = Xh(X)$ and h divides $f_1, \ldots, f_n$, so we can write $f_j = r_j h$ ($j = 1, 2, \ldots, n$) with suitable $r_j \in \mathbf{Z}[X]$. This leads to

$$X\mathbf{Z}[X] = p \sum_{j=1}^{n} r_j \mathbf{Z}[X] + X \sum_{j=1}^{n} r_j \mathbf{Z}[X],$$

which shows that all r_j's are divisible by X, but this is not possible, since dividing the last equation by X we obtain $\mathbf{Z}[X] \subset M$, which is definitely false.

b) $g(0) \neq 0$, i.e. X does not divide g. In this case all f_j's must be divisible by X, so write $f_j(X) = gr_j(X)$ with suitable $r_j \in \mathbf{Z}[X]$. Thus

$$g\mathbf{Z}[X] = g(p \sum_{j=1}^{n} r_j \mathbf{Z}[X] + X \sum_{j=1}^{n} r_j \mathbf{Z}[X]),$$

and after dividing by g we again obtain $\mathbf{Z}[X] \subset M$, a contradiction.

9. Now we prove that $Int(\mathbf{Z})$ is Prüfer, a result of D.BRIZOLIS [79], who actually obtained this assertion for a more general class of rings $Int(R)$, assuming R to be a Dedekind domain with finite norm property and satisfying additionally the following conditions:

(a) *R is of zero characteristics,*

(b) *For every non-constant polynomial $f \in R[X]$ the congruence*

$$f(x) \equiv 0 \pmod{P}$$

is solvable for infinitely many prime ideals P of R.

It has been shown later (J.-L.CHABERT [87], D.L.MCQUILLAN [85A]) that the conditions (a), (b) are redundant and in fact $Int(R)$ is a Prüfer domain for a Dedekind domain R if and only if R has the finite norm property. J.-L.CHABERT [87] proved that if R is a Noetherian domain then $Int(R)$ can be a Prüfer domain only if R is a Dedekind domain. A description of non-Noetherian domains R for which $Int(R)$ is a Prüfer domain is not known (**PROBLEM X**). Some necessary conditions has been established by D.L.MCQUILLAN [85a] and J.-L.CHABERT [91] .

THEOREM 7.8. (D.BRIZOLIS [79]) *The ring $Int(\mathbf{Z})$ is a Prüfer domain.*

PROOF: Let M be a maximal ideal of $Int(\mathbf{Z})$. Theorem 7.4 shows that there exists a prime p and a p-adic integer c such that

$$M = \{f \in Int(\mathbf{Z}) : f(c) \in p\mathbf{Z}_p\}.$$

Denote by L the localization $Int(\mathbf{Z})_M$, let f, g be two non-zero elements of L and put $r = f/g$. We have to show that either r or $1/r$ belongs to L. We can write $r = F/G$, where $F, G \in \mathbf{Z}[X]$ and F, G do not have a common factor. If either $F(c) \notin p\mathbf{Z}_p$ or $G(c) \notin p\mathbf{Z}_p$ then obviously r or $1/r$ lies in L and we are ready.

Assume thus that $F(c) \in p\mathbf{Z}_p$ and $G(c) \in p\mathbf{Z}_p$. Let v_p be the p-adic absolute value. We shall construct two sequences $f_1, f_2, \dots$ and $g_1, g_2, \dots$ of elements of $Int(\mathbf{Z})$ such that for all i one has $f_i/g_i = r$ and moreover if for some $i = 1, 2, \dots$ one has

$$v_p(f_i(c)) < 1 \quad \text{and} \quad v_p(g_i(c)) < 1$$

then

$$1 \geq v_p(g_{i+1}(c)) > v_p(g_i(c)).$$

Put $f_1 = F$, $g_1 = G$ and if the polynomials f_i, g_i are already defined, then put

$$f_{i+1} = \frac{(f_i^p - f_i)(g_i^{p-1} - 1)}{p}$$

and

$$g_{i+1} = \frac{(g_i^p - g_i)(f_i^{p-1} - 1)}{p}$$

Obviously $\dfrac{f_{i+1}}{g_{i+1}} = r$ and the little Fermat's theorem shows that both f_{i+1} and g_{i+1} lie in $Int(R)$. If $v_p(f_i(c)) < 1$ and $v_p(g_i(c)) < 1$ then we get

$$v_p(f_i^{p-1}(c)) \le v_p(f_i(c)) < 1$$

and

$$v_p(g_i^p(c)) < v_p(g_i(c)) < 1,$$

giving

$$v_p(f_i^{p-1}(c) - 1) = 1$$

and

$$v_p(g_i^p(c) - g_i(c)) = v_p(g_i(c)),$$

which leads to

$$v_p(g_{i+1}(c)) = \frac{v_p(g_i(c))}{v_p(p)} > v_p(g_i(c)),$$

as asserted.

Since the set of values of v_p is discrete, for a certain i we must have either $v_p(g_i(c)) \ge 1$ or $v_p(f_i(c)) \ge 1$. In the first case we get $r \in L$ and in the second $1/r \in L$. $\square$

A related question has been answered by D.L.McQuillan [85c] who showed that if R is a Prüfer domain and A a finite subset of it, then the ring $Int(A, R)$ is Prüfer.

10. It is a classical result that in a Dedekind domain every non-principal ideal can be generated by two elements (see e.g. [EATAN], Corollary 5 to Proposition 1.6). More generally, this property has every invertible ideal in a commutative ring with unit element, provided it is contained in finitely many maximal ideals (E.Matlis [66]). For a description of domains in which every ideal has two generators see E.Matlis [70]. It has been established by R.C.Heitmann [76] that if R is a Prüfer domain whose Krull dimension equals m, then every finitely generated ideal of R can be generated by a set of at most $m + 1$ elements. (The case $m = 1$ is due to J.Sally, W.Vasconcelos [74]). From Corollary 2 to Theorem 7.4 it follows that every finitely generated ideal of $Int(\mathbf{Z})$ can be generated by three elements. R.Gilmer and W.W.Smith [83] showed that one can replace here the number 3 by 2, and this has been extended to $Int(R)$ for any Dedekind domain R with finite norm property by D.L.McQuillan [85a]. (Cf. J.-L.Chabert [87]). (For the case when R is the ring of integers in an algebraic number field this has been shown earlier by D.E.Rush [85]).

Note that H.W.SCHÜLTING [79] showed that not all two-dimensional Prüfer domains have this property.

It has been shown by R.GILMER and W.W.SMITH [83] that if $f(X) = c$ is a constant polynomial in $Int(\mathbf{Z})$ and I is a finitely generated ideal of $Int(\mathbf{Z})$ containing f, then with a suitable $g \in I$ one has $fInt(\mathbf{Z}) + gInt(\mathbf{Z}) = I$. (Every such f is said to be a *strong 2-generator*). This result has been extended to all Dedekind domains with finite norm property by J.-L.CHABERT [87]. Later R.GILMER and W.W.SMITH [85] showed that not every element of $Int(\mathbf{Z})$ is a strong 2-generator, in fact they proved that if $f \in Int(\mathbf{Z})$ is non-constant, irreducible in $\mathbf{Q}[X]$, and such that the quotient ring $Int(\mathbf{Z})/(f\mathbf{Q}[X] \cap S)$ is not a principal ideal domain, then f cannot be a strong 2-generator. One can take here e.g. $f(X) = X^2 - d$, provided the class-group of the field $\mathbf{Q}(\sqrt{d})$ contains elements of order exceeding 2. (Cf. J.BREWER, L.KLINGLER [91]). No description of strong 2-generators in $Int(\mathbf{Z})$ is known (**PROBLEM XI**).

11. The *Picard group* (the factor group of the group of invertible ideals by the group of principal ideals) of $Int(\mathbf{Z})$ has been determined in R.GILMER, W.HEINZER, D.LANTZ, W.SMITH [90]. It turned out to be isomorphic with the free abelian group having countably many free generators.

Exercises

1. Let p be a rational prime, $c \in \mathbf{Z}_p \setminus \mathbf{Z}$ and let $M = M(p, c)$ be the corresponding maximal ideal of $Int(Z)$ (as defined in Theorem 7.4).

(i) Show that for every $n \in \mathbf{Z}$ there exists $f \in M$ with $f(n) = 1$.

(ii) Prove that M is not finitely generated.

2. Let R be a domain.

(i) Prove that if R is not a field, then the Krull dimension of $Int(R)$ is ≥ 2.

(ii) Prove that $Int(R)$ is a Dedekind domain if and only if R is a field.

(iii) Show that if $Int(R)$ is Noetherian, so is R.

(iv) Prove that $Int(R)$ is integrally closed if and only if R is integrally closed.

3. (D.BRIZOLIS [75]) Let K be a field. Prove that the ring $K[X_1, \ldots, X_n]$ has the Skolem property if and only if K is algebraically closed.

4. (J.-L.CHABERT [71]) Let A be a domain and L it field of quotients. A rational fraction $P/Q \in L(X)$, with $P, Q \in K[X]$ is called *normalized* if $\deg P < \deg Q$, $Q(0) = 1$ and P, Q are relatively prime in $L[X]$. Show that if R is a domain with quotient field K, $A = R[T]$ and $P(X, T)/Q(X, T) \in K(X)(T)$ is a normalized fraction then for all $r \in R$ with at most finitely many exceptions the fraction $P(r, T)/Q(r, T)$ is normalized.

5. (J.-L.CHABERT [71]) A domain R with field of quotients K is called a *Fatou ring* (B.BENZAGHOU [70]) if every normalized fraction $P/Q \in K(X)$

whose power series has coefficients in R satisfies $P, Q \in R[X]$. Prove that if R is a Fatou ring, then $Int(R)$ has this property too.

VIII. D-rings

1. In this section we shall consider a ring property which is related to mappings induced by rational functions. Let R be a domain and denote by $S(R)$ the ring of all rational functions $f \in R(X)$ whose values at these points of R which are not poles of f lie in R. Clearly $Int(R) \subset S(R)$ and it turns out that $Int(R) \neq S(R)$ holds if and only if there are non-constant polynomials in $R[X]$ which map R into $U(R)$, its *group of units*, i.e. the group of all invertible elements of R. This forms a part of the following theorem established by H.GUNJI, D.L.McQUILLAN [75] and D.BRIZOLIS [75]:

THEOREM 8.1. *Let R be a domain and K its field of fractions. For any polynomial $f \in R[X]$ denote by $I(f)$ the set of all non-zero prime ideals P of R for which the congruence $f(x) \equiv 0 \pmod{P}$ is solvable in R. The following properties are equivalent:*

(a) *If f, g are polynomials over R such that for almost all $r \in R$ (i.e. for all $r \in R$ with at most finitely many exceptions) $f(r)$ divides $g(r)$ then the ratio $g(X)/f(X)$ is a polynomial with coefficients in K,*

(b) *Every polynomial over R which for almost all $r \in R$ satisfies $f(r) \in U(R)$ must be a constant,*

(c) *For any non-constant polynomial $f \in R[X]$ the set $I(f)$ is non-empty,*

(d) *For any non-constant polynomial $f \in R[X]$ and any nonzero $c \in R$ the set $I(f) \setminus I(c)$ is infinite.*

PROOF: (a) $\Longrightarrow$ (b). If f maps R in $U(R)$ then for almost all $a \in R$, $f(a)$ divides 1, thus by (a) f divides 1 in $K[X]$ and so must be a constant.

(b) $\Longrightarrow$ (c). If $I(f)$ is empty, then for every $r \in R$ one has $f(r) \in U(R)$, contradicting (b).

(c) $\Longrightarrow$ (d). Assume $I(f) \setminus I(c)$ to be finite, and denote by m a non-zero element of the product of all prime ideals lying in $I(f) \setminus I(c)$.

First consider the case $f(0) = 0$. In this case for every $a \in R$ one has $f(a) \in aR$ thus $I(f)$ consists of all prime ideals of R and $I(c)$ contains all but finitely many such ideals. Since cm is non-zero and lies in every prime ideal of R, the polynomial $g(X) = 1 + cmX$ maps R in $U(R)$, thus $I(g)$ is empty, contradicting (c).

Now let $f(0) = d \neq 0$. All coefficients of the polynomial $f(cdX)$ being divisible by d, we may write $f(cdX) = dg(X)$ with a suitable polynomial $g \in R[X]$. Since the free term of g equals 1 and the remaining coefficients are divisible by c thus for every $a \in R$ we get $g(a) \equiv 1 \pmod{P}$ for every prime ideal $P \in I(c)$. This shows that $I(g)$ and $I(c)$ are disjoint. In view of $I(g) \subset I(f)$ we get

$$I(g) \subset I(f) \setminus I(c)$$

and it remains to establish the infiniteness of $I(g)$. But this is easy. If $I(g)$ were finite and b would be a non-zero element lying in the product of all members of $I(g)$, then the polynomial $h(X) = g(bX)$ would have its free term equal to 1 whereas all other its coefficients would be divisible by b. Thus for all $x \in R$ and all prime ideals $P \in I(g)$ we would have $h(x) \equiv 1 \pmod{P}$, but $I(h) \subset I(g)$, and so $I(h)$ would be empty, contradicting (c).

(d) $\implies$ (a) Assume that the condition (d) holds and that for almost all $r \in R$, $f(r)$ divides $g(r)$. Without restricting the generality we may assume that the polynomials f and g are relatively prime in $K[X]$ and thus with suitable polynomials $A, B \in R[X]$ and a $c \in R$ we can write

$$A(X)f(X) + B(X)g(X) = c.$$

This shows that for almost all $r \in R$, $f(r)$ divides c. Assume now that f is non-constant. If P is a prime ideal belonging to $I(f) \setminus I(c)$, then with a suitable $r \in R$ we have $f(r) \equiv 0 \pmod{P}$. Replacing, if necessary, r by a suitable element congruent to $r \pmod{P}$ we may assume that $f(r)$ divides c, but then c, being a multiple of $f(r)$, would belong to P, contradiction. Hence f must be constant. $\square$

COROLLARY 1. *If R is a domain, then the condition (a) of the theorem is equivalent to $S(R) = Int(R)$.*

PROOF: Obviously (a) implies $S(R) = Int(R)$. To prove the converse assume $S(R) = Int(R)$, let f, g be polynomials with coefficients in R and let $A = \{r_1, r_2, \ldots, r_t\}$ be a finite subset of R with the property that for every $r \in R$ outside A one has $f(r) \mid g(r)$. Let $r_1, r_2, \ldots, r_s$ be those members of A which are not zeros of f (if any), and put $c = f(r_1) \cdots f(r_s)$. In case $s = 0$ put $c = 1$. Then for all elements $r \in R$ which are not zeros of f we have $f(r) \mid cg(r)$ and our assumption implies $cg(X)/f(X) \in K[X]$. So $g(X)/f(X)$ must be a polynomial. $\square$

2. Domains which satisfy the equivalent conditions of Theorem 8.1 have been called *D-rings* by H.GUNJI and D.L.MCQUILLAN [75] and *rings with property S* by D.BRIZOLIS [75] .

The paper of H.Gunji and D.L.McQuillan contains a study of these rings. It is proved there that all rings consisting of algebraic integers as well as all rings which are not fields and whose group of units is finitely generated are *D*-rings. These authors showed also that if R is a *D*-ring and S is a ring containing R, which is either finitely generated or integral over R then S is also a *D*-ring.

COROLLARY 2. *If R is a Dedekind domain of zero characteristics with field of quotients K, then R is a D-ring if and only if in every finite extension L/K*

infinitely many prime ideals of R split completely. In particular the ring of all integers of an algebraic number field is a D-ring.

PROOF: Kummer's theorem (see e.g. [EATAN], Theorem 4.12) shows that infinitely many prime ideals split in L/K if and only if the condition (c) of the theorem holds. □

COROLLARY 3. (D.BRIZOLIS [74]. The case of the ring of rational integers is due to D.A.LIND [71]) *The ring of all integers of an algebraic number field is a D-ring and in particular* **Z** *is a D-ring.*

PROOF: This is a direct consequence of the preceding corollary since by a well-known theorem infinitely many prime ideals split in every finite extension of the rationals (see e.g. [EATAN], Theorem 4.14). We give now a direct elementary argument, due to D.BRIZOLIS [74]:

Let R be the ring of integers of an algebraic number field, let f, g be polynomials over R and assume that for almost all $r \in R$ we have $f(r) \mid g(r)$. For any polynomial $F \in R[X]$ denote by NF its norm i.e. the product of all polynomials conjugated with F over $\mathbf{Q}$. Observe that for rational a we have $NF(a) = N_{K/\mathbf{Q}}(F(a))$. This fact and our assumptions imply now that for sufficiently large rational integral a one has

$$\left| \frac{Ng(a)}{Nf(a)} \right| = \left| N_{K/\mathbf{Q}} \left(\frac{g(a)}{f(a)} \right) \right| \geq 1,$$

because $g(a)/f(a)$ is a non-zero algebraic integer. Hence we must have deg $f \leq$ deg g, since otherwise the left-hand side of the preceding inequality would tend to zero for a tending to infinity.

Now we proceed by induction. The assertion being evident in the case of constant polynomials f and g we assume that it holds for all polynomials of degrees not exceeding $n - 1$. Let now

$$g(X) = AX^n + \cdots \quad \text{and} \quad f(X) = BX^m + \cdots,$$

with $AB \neq 0$ and $m \leq n$. If we put $h(X) = g(X) - Af(X)X^{n-m}/B$, then in view of

$$Bh(X) = Bg(X) - Af(X)X^{n-m}$$

we obtain that for almost all $r \in R$ one has $f(a) \mid Bh(a)$ and since the degree of $Bh(X)$ is smaller than n the inductional assumption shows that Bh/f is a polynomial, hence Bg/f must be a a polynomial and the same applies to g/f. □

COROLLARY 4. (H.GUNJI, D.L.McQUILLAN [75]) *The ring of polynomials in any number of variables over a domain is a D-ring.*

PROOF: It suffices to consider the case of one variable. In this case observe that the composition of two polynomials of positive degrees has a positive degree and hence the condition (b) is satisfied. □

Obviously no field can be a D-ring. The next result shows that there other simple examples:

COROLLARY 5. (H.GUNJI, D.L.McQUILLAN [75]) *Let R be the smallest ring containing the rational integers and all numbers $1/p$, where p runs over primes not congruent to 3 (mod 4). Then R is not a D-ring.*

PROOF: Using the observation that no prime $p \equiv 3 \pmod 4$ can divide the sum of two relatively prime integral squares, one sees easily that the polynomial $X^2 + 1$ maps R into its group of units. □

It has been noted by D.A.LIND [71] that no localization $\mathbf{Z}_{(p)}$ of $\mathbf{Z}$ with respect to a prime p can be a D-ring. In fact, the rational function $X/(X^2 + p)$ maps $\mathbf{Z}_{(p)}$ into $\mathbf{Z}_{(p)}$. The same argument applies to localizations of Dedekind domains and leads hence to a large class of Dedekind domains which are not D-rings. However this procedure does not exhaust all cases of such domains. A.LOPER [88] constructed examples of Dedekind domains R with class-group $H(R) = C_2$, which are not localizations of Dedekind domains but are not D-rings. His examples are localizations of polynomial rings in one variable over a Dedekind domain. He showed that in this case the classgroup must be a torsion group and the degree of any unit-valued polynomial, which exists according to Theorem 8.1 must be a multiple of the exponent of the class-group $H(R)$. It is not known whether one can find such examples with a larger class-group (**PROBLEM XII**), nor whether they always must be localizations (**PROBLEM XIII**).

If R is a non-trivial valuation ring of a field K and M is its maximal ideal, then for any non-zero $a \in M$ the rational function $f(X) = 1/(1 + aX)$ lies in $S(R) \setminus Int(R)$, hence R is not a D-ring. It has been established by A.PRESTEL,C.C.RIPOLI [91] that in this case one has

$$S(R) := \left\{ \frac{f(X)}{1 + cg(X)} : f, g \in Int(R), c \in M \right\}$$

if and only if the completion of K is either locally compact or algebraically closed. The case of $R = \mathbf{Z}_p$, the ring of p-adic integers has been treated earlier by S.KOCHEN [69] and P.ROQUETTE [71] . Cf. P.-J.CAHEN [78].

3. The truth of the analogue of condition (a) in Theorem 8.1 for polynomials in several variables in rings of integers of algebraic number fields has been obtained by D.J.LEWIS and P.MORTON [81]. For the ring of rational integers this belongs to the folklore (see e.g. W.NARKIEWICZ [66]), however in this case much more is true. The following result has been established by Lewis and Morton in the quoted paper:

Let $m_1, \ldots, m_n$ be pairwise relatively prime rational integers, and let f, g be two polynomials in n variables with rational integral coefficients. If for $k = 1, 2, \ldots$ one has

$$f(m_1^k, \ldots, m_n^k) \mid g(m_1^k, \ldots, m_n^k),$$

provided the left-hand side does not vanish, then g/f is a polynomial. The same holds for the ring of integers in a real algebraic number field.

4. We conclude this section with a presentation of certain properties related to the Skolem property and D-rings. One says that a ring A, satisfying $R[X] \subset A \subset Int(R)$, has the *strong Skolem property* (D.BRIZOLIS [75], J.-L.CHABERT [82]) if for every pair I, J of finitely generated ideals of D the equalities $I(r) = J(r)$ (for all $r \in R$) imply $I = J$. This notion can be traced back to to the paper of T.SKOLEM [37b], the main result of which shows that $Int(\mathbf{Z})$ has the strong Skolem property.

If this implication holds for all finitely generated ideals I, J which satisfy additionally $I \cap R \neq 0$ and $J \cap R \neq 0$ then D is said to have the *strong Hilbert property* (D.L.MCQUILLAN [85a]). For Dedekind domains this property is equivalent to the finite norm property. For interesting examples see R.GILMER [90] .

Finally, if in the definition of the Skolem property one considers only finitely generated ideals having non-zero intersection with R, then one obtains the *Hilbert property* (D.L.MCQUILLAN [85a]). Noetherian domains R (which are not fields) for which $Int(R)$ has the Hilbert property have been characterized by J.-L.CHABERT [88] as those which have the following two properties:

(i) *Every non-zero prime ideal of R is the intersection of maximal ideals,*

(ii) *If M is a maximal ideal of R such that the field A/M is not algebraically closed, then M is of height one and A/M is finite.*

Moreover he showed that if R is a domain, then $Int(R)$ has the strong Skolem property if and only if R is a D-ring having the strong Hilbert property.

We prove now a result of D.BRIZOLIS [79] which implies in particular that the ring $Int(\mathbf{Z})$ has the strong Skolem property:

THEOREM 8.2. *Let R be a Dedekind domain with quotient field K. If R is a D-ring and $Int(R)$ is a Prüfer domain, then $Int(R)$ has the strong Skolem property.*

PROOF: Let I, J be two finitely generated ideals of $Int(R)$ and assume that for all $r \in R$ one has $I(r) = J(r)$. Since $Int(R)$ is a Prüfer domain, the ideal J is invertible hence there exists a fractional ideal J^{-1} satisfying $JJ^{-1} \subset Int(R)$. Let φ be a rational function lying in IJ^{-1}. We can write

$$\varphi = \sum_{i=1}^{N} f_i g_i,$$

with $f_i \in I$ and $g_i \in J^{-1} \subset K(X)$ for $i = 1, 2, \ldots, N$. Denote by $\mathcal{X}$ the finite set of all poles of $\prod_{i=1}^{N} g_i$.

LEMMA 8.3. *For $i = 1, 2, \ldots, N$ and $r \in R \setminus \mathcal{X}$ one has*

$$g_i(r) \in I(r)^{-1}.$$

PROOF: Let s be a non-zero element of $J(r)$. Thus for a suitable $h \in J$ we have $s = h(r)$. Since for $i = 1, 2, \ldots, N$ we have $hg_i \in JJ^{-1} \subset Int(\mathbf{Z})$ we obtain

$sg_i(r) = h(r)g_i(r) = hg_i(r) \in \mathbf{Z}$ and with a suitable $m_i \in \mathbf{Z}$ we get

$$g_i(r) = m_i/h(r) \in J(r)^{-1} = I(r)^{-1}. \quad \square$$

The lemma implies that for $r \in R \setminus \mathcal{X}$ one has $\varphi(r) \in \mathbf{Z}$ and since R is a D-ring, Theorem 8.1 shows that $\varphi \in Int(R)$. This establishes $IJ^{-1} \subset Int(R)$ thus $I \subset J$ and by the symmetry of our assumption we get also $J \subset I$. leading finally to $I = J$. $\square$

COROLLARY. (T. SKOLEM [37b] $Int(\mathbf{Z})$ has the strong Skolem property.

PROOF: We know from Theorem 7.8 that $Int(R)$ is a Prüfer domain and Corollary 3 to Theorem 8.1 shows that $\mathbf{Z}$ is a D-ring, so we may apply the theorem just proved. $\square$

Exercises

1. Let K be a field.

(i) Prove that $K[X]$ has the strong Skolem property if and only if K is algebraically closed.

(ii) Show that $K[X]$ has the strong Hilbert property.

2. Give an example of a domain R for which $Int(R)$ has the strong Hilbert property but not the Skolem property.

3. (T. SKOLEM [37b]) Let R be the ring of all integral-valued polynomials in n variables over $\mathbf{Q}$ and for every ideal I of R and $a_1, \ldots, a_n \in \mathbf{Z}$ put

$$I(a_1, \ldots, a_n) = \{f(a_1, \ldots, a_n) : f \in I\}.$$

Show that if I, J are finitely generated ideals of R and for all $a_1, \ldots, a_n \in \mathbf{Z}$ one has $I(a_1, \ldots, a_n) \subset J(a_1, \ldots, a_n)$, then $I \subset J$.

4. (J.-L.CHABERT [88]) Let R be a domain. Prove that $Int(R)$ has the Skolem property if and only if it has the Hilbert property and R is a D-ring.

5. (J.-L.CHABERT [88]) Let K be a field and $R = K[X]$. Show that $Int(R)$ has the strong Skolem property if and only if K is finite.

6. Let R be a ring with non-trivial *Jacobson radical* (i.e. intersection of all maximal ideals). Show that R is not a D-ring.

7. Let $f, g \in \mathbf{Z}[X_1, \ldots, X_n]$ and assume that for all $a_1, a_2, \ldots, a_n \in \mathbf{Z}$ with at most finitely many exceptions one has

$$f(a_1, a_2, \ldots, a_n)|g(a_1, a_2, \ldots, a_n).$$

Prove that the ratio f/g is a polynomial over $\mathbf{Q}$.

Part B

Fully invariant sets for polynomial mappings

IX. The properties (P) and (SP)

1. It is an easy observation that if f is a polynomial which maps the rational number field $\mathbf{Q}$ onto itself, then f must be linear. In fact, if there is a non-linear polynomial with this property then there exists also such a polynomial with integral coefficients, say

$$g(X) = a_N X^N + \cdots + a_0.$$

We may assume $a_N > 0$, thus g increases in $[a, \infty)$ if a is chosen sufficiently large. Let $x_n = p_n/q_n$ $((p_n, q_n) = 1)$ be for $n = 1, 2, \ldots$ a positive rational number with $g(x_n) = n$. For sufficiently large n we have $x_{n+1} > x_n$ and since q_n must be a divisor of a_N we obtain $x_{n+1} - x_n \geq 1/a_N$ and thus $1 = f(x_{n+1}) - f(x_n)$ tends to infinity, contradiction.

If F is a mapping of a set Ω into itself, then a subset X of Ω is called *invariant* (with respect to F), provided $F(X) \subset X$ and it is called *completely invariant* provided $F(X) = X$ and $F^{-1}(X) = X$ holds. The last property has been introduced by D.FATOU [19] who started an intensive study of this notion in the case of maps of the complex plane. For an introduction into that theory see [IRF].

We shall consider here a weaker property, viz. $F(X) = X$, and sets posessing it we shall call, for lack of a better name, *fully invariant sets* . The trivial observation made above shows that $\mathbf{Q}$ is not a fully invariant set for a non-linear polynomial mapping. It turns out that in this assertion one may replace the field $\mathbf{Q}$ by any of its infinite subsets, and thus a non-linear polynomial map $\mathbf{Q} \longrightarrow \mathbf{Q}$ cannot have infinite fully invariant sets. We shall obtain this result as a special case of a more general statement dealing with mappings defined by systems of polynomials in several variables.

2. Let Ω be an arbitrary set and let $X \subset \Omega$ be fully invariant with respect to a mapping $F : \Omega \longrightarrow \Omega$. If $x \in X$ and we put $x_0 = x$, $x_{n+1} = F(x_n)$ for

$n = 0, 1, 2, \ldots$ then clearly $x_0, x_1, \ldots \in X$. If for some $j > 0$ we have $x_j = x_0$ then the set $\{x_0, x_1, \ldots, x_{j-1}\}$ is fully invariant and we shall call it a *finite F-cycle*. If $x \in X$ is not contained in any finite F-cycle, then for $n = -1, -2, \ldots$ we may choose $x_n \in X$ satisfying $F(x_n) = x_{n+1}$ with elements $x_{-1}, x_{-2}, \ldots$ distinct. The sequence $\{x_n\}_{-\infty}^{\infty}$ will be called an *infinite F-cycle*. It also forms a fully invariant set.

Note that two finite F-cycles either coincide or are disjoint, and if $\{x_n\}$, $\{y_n\}$ are two infinite F-cycles which have an element in common, say $x_n = y_m$, then for $j = 0, 1, 2, \ldots$ one has $x_{n+j} = y_{m+j}$. It may happen that an infinite cycle contains a finite cycle as a subset. In fact, an example arises by taking $\Omega = \mathbf{C}$ and putting $F(z) = z^2 - 1$. Then the numbers 0 and -1 form a finite cycle contained in the infinite cycle defined as follows: put $x_{-1} = 1$, for $n = -2, -3, \ldots$ define x_n by

$$x_n = \sqrt{1 + x_{n+1}},$$

and put $x_{2n} = 0$, $x_{2n+1} = -1$ for $n = 1, 2, \ldots$.

An element $x \in \Omega$ is called *cyclic with respect to F* if it lies in a finite F-cycle. The set of all cyclic elements with respect to F will be denoted by $Cycl_\Omega(F)$ or simply $Cycl(F)$.

3. We need two definitions dealing with polynomial mappings in several variables defined over arbitrary fields:

Let K be a field and let $f_1, \ldots, f_n$ be polynomials in n variables with coefficients in K. The mapping

$$F : K^n \longrightarrow K^n$$

defined by

$$[x_1, \ldots, x_n] \mapsto [f_1(x_1, \ldots, x_n), \ldots, f_n(x_1, \ldots, x_n)]$$

is called *admissible* provided none of the polynomials $f_1, \ldots, f_n$ is linear and their leading forms do not have any non-trivial common zero in the algebraic closure of K. (Recall that the *leading form* of a polynomial

$$f(X_1, \ldots, X_n) = \sum_{i_1, \ldots, i_n} a_{i_1, \ldots, i_n} X_1^{i_1} \cdots X_n^{i_n}$$

is defined as the sum of non-zero monomials $a_{i_1, \ldots, i_n} X_1^{i_1} \cdots X_n^{i_n}$ with maximal sum $i_1 + i_2 + \cdots + i_n$).

A field K is said to have the *property* **(SP)** if for every n and for every admissible polynomial mapping

$$F : K^n \to K^n$$

there are no infinite fully invariant sets, i.e. the conditions

$$X \subset K^n, \quad F(X) = X$$

imply the finiteness of X.

If this implication holds in the case $n = 1$ then we say that it has the *property* **(P)**. Clearly **(SP)** implies **(P)** but it is not known whether the converse implication holds **(PROBLEM XIV)**.

Note that if a field has the property **(SP)** then admissible polynomial mappings cannot have infinite cycles.

The examples, in which $K = \mathbf{Q}$, $n = 2$ and $X = \{[a,a] : a \in \mathbf{Q}\}$,

$$f_1(X_1, X_2) = (X_1 - X_2)g_1(X_1, X_2) + X_1$$
$$f_2(X_1, X_2) = (X_1 - X_2)g_2(X_1, X_2) + X_2$$

and

$$f_1(X_1, X_2) = X_1^2 - X_2^2 + X_1$$
$$f_2(X_1, X_2) = 2X_1 - X_2$$

(where g_1, g_2 are arbitrary forms) explain the necessity of considering only admissible mappings in the definition of **(SP)**.

It is obvious that if a field has one of the properties **(P)** or **(SP)** then every its subfield does the same. We shall later prove that the property **(SP)** has also other nice properties: it is preserved by arbitrary finite extensions (Theorem 9.4) as well by all purely transcendental extensions (Theorem 10.13), and so in particular all global fields and, more generally, all finitely generated fields have this property. About the property **(P)** it is known that it is preserved by purely transcendental extensions (Theorem 10.10) but it is an open question whether it is preserved also by finite extensions **(PROBLEM XV)**. A constructive description of fields having the property **(P)** or **(SP)** is unknown **(PROBLEM XVI)**. The following question, concerning a rather particular case of this problem is also open:

PROBLEM XVII: *Does the field generated by all square roots of primes have the property* **(P)** *or* **(SP)**?

4. We start with a simple observation showing that in dealing with the property **(SP)** it is sufficient to consider polynomial mappings which are defined by a system of homogeneous polynomials:

THEOREM 9.1. *Let K be a field and assume that for every n and for every admissible polynomial map*

$$F : K^n \longrightarrow K^n$$

defined by a system of homogeneous forms and for every $X \subset K^n$ the condition

$$F(X) = X$$

implies the finiteness of X. Then the field K has the property **(SP)**.

PROOF: Let $F : K^n \longrightarrow K^n$ be defined by an admissible system $F_1, \ldots, F_n$ of polynomials and for $i = 1, \ldots, n$ write

$$F_i = \sum_{j=0}^{n_i} G_i^{(j)}$$

where $G_i^{(j)}$ is a form of degree j. If we now put for $k = 1, 2, \ldots, n$

$$H_k(X_1, \ldots, X_{n+1}) = \sum_{j=0}^{n_i} X_{n+1}^j G_k^{(n_i - j)}(X_1, \ldots, X_n)$$

and

$$H_{n+1}(X_1, \ldots, X_{n+1}) = X_{n+1}^2,$$

then one sees immediately, that the map $F^* : K^{n+1} \longrightarrow K^{n+1}$ defined by these forms is admissible and if X is a subset of K such that $F(X) = X$ then the set

$$Y = \{[y_1, \ldots, y_n, 1] : [y_1, \ldots, y_n] \in X\} \subset K^{n+1}$$

satisfies $F^*(Y) = Y$, thus is finite and the finiteness of X follows. $\square$

5. Our main tool in this and in the next sections will be the following elementary lemma, dealing with mappings of an arbitrary set, without any algebraic structure. We state it in two variants, the second of which will be easier to apply in the next section.

LEMMA 9.2. (a) (W.NARKIEWICZ [62]) *Let X be a non-empty set and*

$$T : X \longrightarrow X$$

a surjective map. Assume that there are two functions f, g defined on X with positive real values, which satisfy the following three conditions:
   (i) *For every $c > 0$ only finitely many elements $x \in X$ can satisfy $f(x) + g(x) \leq c$,*
   (ii) *There is a constant C such that if $f(x) > C$, then $f(Tx) > f(x)$,*
   (iii) *For every fixed B there is a constant $C_1(B)$ such that if $f(x) \leq B$ and $g(x) > C_1(B)$, then $g(Tx) > g(x)$.*
   Then X is finite.

   (b) (F.HALTER-KOCH, W.NARKIEWICZ [92a]) *Let X be a non-empty set and*

$$T : X \longrightarrow X$$

a surjective map. Assume that there is a real-valued function f defined on X and a positive constant C which satisfy the following conditions:
   (i) *The set $f(X)$ is discrete,*
   (ii) *If $f(x) \geq C$ then $f(Tx) > f(x)$,*
   (iii) *The set*

$$X_0 = \{x \in X : f(x) < C\}$$

is finite.

Then X is finite and $X = X_0$.

PROOF: (a) Since T is surjective we may for any $x_0 \in X$ construct a sequence $x_0, x_1, \ldots$ of elements of X satisfying

$$T(x_i) = x_{i-1} \quad (i = 1, 2, \ldots).$$

If for some $k \geq 1$ one has $f(x_k) > C$ and k is the first index with this property, then (ii) implies $f(x_k) < f(x_{k-1})$, a contradiction. It follows that $f(x_i) \leq C$ holds for all i. The same argument shows that the sequence $\{g(x_i)\}$ is bounded by $\max(C_1(C))$ and

$$f(x_0) + g(x_0) \leq C + C_1(C)$$

follows. Since x_0 was an arbitrary element of X, the application of (i) leads to the finiteness of X.

(b) If $x \notin X_0$, then $f(Tx) > f(x) \geq C$, hence $Tx \notin X_0$. Since T is surjective, we get $X_0 \subset T(X_0)$ and as X_0 is finite, $T(X_0) = X_0$ follows. If $X \neq X_0$, $x \in X \setminus X_0$ is chosen to have $f(x)$ minimal and $y \in X$ satisfies $T(y) = x$, then $y \notin X_0$, so $f(y) \geq C$ and (i) leads to $f(x) = f(Ty) > f(y)$, contradicting the choice of x. Thus $X = X_0$ and the assertion follows. $\square$

COROLLARY. Let X be a non-empty set and

$$T : X \longrightarrow X$$

a surjective map. If there exists a function F defined on X with positive real values such that for every $c > 0$ the inequality $F(x) \leq c$ is satisfied only by finitely many elements $x \in X$ and there is a constant C such that if $f(x) > C$, then $f(Tx) > f(x)$, then the set X is finite.

PROOF: Apply either part (a) of the lemma with $f = F$ and $g = 1$ or part (b), since the assumptions imply the discreteness of $F(X)$. $\square$

6. Consider now the field $\mathbf{Q}$ of rational numbers.

THEOREM 9.3. The field $\mathbf{Q}$ has the property (SP).

PROOF: Let an admissible map $F : \mathbf{Q}^n \longrightarrow \mathbf{Q}^n$ be defined by the polynomials

$$f_1(X_1, \ldots, X_n), \ldots, f_n(X_1, \ldots, X_n)$$

with rational coefficients. In view of the preceding theorem we may assume that all f_i's are homogeneous. Denote by r_i the degree of f_i, put

$$R = \max\{r_1, \ldots, r_n\}, \quad r = \min\{r_1, \ldots, r_n\}$$

and write

$$f_i(X_1, \ldots, X_n) = \frac{F_i(X_1, \ldots, X_n)}{D} \quad (i = 1, \ldots, n)$$

where D is a positive integer and $F_1, \ldots, F_n$ are forms with rational integral coefficients.

Every element $\boldsymbol{\xi} \in \mathbf{Q}^n$ can be uniquely written in the form

$$\boldsymbol{\xi} = \left[\frac{p_1}{q}, \ldots, \frac{p_n}{q}\right],$$

where $p_1, \ldots, p_n, q$ are integers, $q > 0$ and $(p_1, \ldots, p_n, q) = 1$. To apply Lemma 9.2 (a) we put

$$f(\boldsymbol{\xi}) = q, \qquad g(\boldsymbol{\xi}) = \max\{|p_i| : i = 1, \ldots, n\}.$$

The condition (i) of the lemma is obviously satisfied. To check the other conditions observe that we have

$$F(\boldsymbol{\xi}) = \left[q^{R-r_1}\frac{F_1(p_1, \ldots, p_n)}{Dq^R}, \ldots, q^{R-r_n}\frac{F_n(p_1, \ldots, p_n)}{Dq^R}\right]$$

and this implies

$$f(F(\boldsymbol{\xi})) \geq \frac{Dq^R}{\mu},$$

where by μ we denoted the maximal divisor of Dq^R dividing $q^{R-r_i}F_i(p_1, \ldots, p_n)$ for all i.

Now we invoke the *Nullstellensatz* of Hilbert (see e.g. [A], §130) to obtain the existence of nonnegative exponents λ_i, rational integers C_i and polynomials $V_j^{(i)}$ with rational, integral coefficients such that

$$q^{R-r}\sum_{j=1}^{n} V_j^{(i)}(X_1, \ldots, X_n)F_j(X_1, \ldots, X_n) = C_i X_i^{\lambda_i} q^{R-r}$$

holds for $i = 1, \ldots, n$.

Putting here $X_i = p_i$ for $i = 1, 2, \ldots, n$ we obtain

$$\mu \mid q^{R-r}(C_1 p_1^{\lambda_1}, \ldots, C_n p_n^{\lambda_n})$$

and this leads to the inequality

$$\mu \leq B_0 q^{R-r},$$

where B_0 is a positive number not depending on $\boldsymbol{\xi}$. Thus we arrive at

$$f(F(\boldsymbol{\xi})) \geq \frac{Dq^r}{B_0} = \frac{D}{B_0}f(\boldsymbol{\xi})^r$$

and in view of $r \geq 2$ the inequality

$$f(F(\boldsymbol{\xi})) > f(\boldsymbol{\xi})$$

results for sufficiently large $f(\boldsymbol{\xi})$.

Thus the condition (ii) of the lemma is satisfied in our case.

Now let $B > 0$ be fixed. To check the last condition of the lemma we shall show that $f(\boldsymbol{\xi}) \leq B$ implies $g(F(\boldsymbol{\xi})) \geq cg(\boldsymbol{\xi})^r$, where c is a positive constant, which does not depend on $\boldsymbol{\xi}$. If this were not true, then there would exist a sequence $\xi_1, \xi_2, \ldots$ satisfying $f(\xi) \leq B$ and

$$\lim_{j \to \infty} \frac{g(F(\xi_j))}{g(\xi_j)^r} = 0.$$

Write

$$\xi_j = \left[\frac{p_1^{(j)}}{q^{(j)}}, \ldots, \frac{p_n^{(j)}}{q^{(j)}} \right]$$

with $p_i^{(j)}$, $q^{(j)} \in \mathbf{Z}$ and observe that by choosing a suitable subsequence we may assume that $q^{(i)} = q \leq B$ is fixed and with a certain i_0 we have

$$g(\xi_j) = |p_{i_0}^{(j)}|$$

for all j and moreover for $i = 1, 2, \ldots, n$ the limits

$$\lambda_i = \lim_{j \to \infty} \frac{p_i^{(j)}}{p_{i_0}^{(j)}}$$

exist. In view of $q \leq B$ we get

$$\max\{|F_i(p_1^{(j)}, \ldots, p_n^{(j)})| : \ i = 1, 2, \ldots, n\} \leq cg(F(\xi))$$

with a certain $c > 0$ and our assumption implies

$$\lim_{j \to \infty} \frac{|F_i(p_1^{(j)}, \ldots, p_n^{(j)})|}{(p_{i_0}^{(j)})^r} = 0,$$

hence

$$\lim_{j \to \infty} \frac{|F_i(p_1^{(j)}, \ldots, p_n^{(j)})|}{(p_{i_0}^{(j)})^{r_i}} = 0,$$

which gives

$$F_i(\lambda_1, \ldots, \lambda_n) = 0 \quad (i = 1, 2, \ldots, n)$$

and thus all λ_i's must vanish, contrary to $|\lambda_{i_0}| = 1$.

All conditions of Lemma 9.2 being satisfied the theorem follows. $\square$

Theorem 9.3 shows that the rational number field has the properties (P) and (SP). Using a similar method one can show that any algebraic number field has it as well (W.NARKIEWICZ [62],[65]). Another proof has been given by P.LIARDET ([70]); it has the advantage of covering all global fields and leading to a stronger assertion (see Theorem 10.7). Later D.J.LEWIS [72] and P.LIARDET [71] showed that every finitely generated field has the property (SP) (see Corollary 3 to Theorem 10.3).

7. Now we shall show that the analogue of Problem XV has for the property (SP) a positive answer:

THEOREM 9.4. (F.HALTER-KOCH, W.NARKIEWICZ [92a]). *Let K be a field having the property* (SP) *and let L/K be a finite extension. Then L has also* *the property* (**SP**).

Proof: If L/K is a purely inseparable extension and $p = \operatorname{char}(K)$ then for a suitable m the map $x \mapsto x^{p^m}$ gives an isomorphism of L with a subfield of K and the assertion follows immediately.

We may thus assume that L/K is a finite separable extension and let n be its degree. Choose a basis $\omega_1, \omega_2, \ldots, \omega_n$ of this extension and note that the determinant of the matrix $[\omega_i^{(j)}]_{i,j=1,\ldots,n}$ (where by $a^{(1)} = a, \ldots, a^{(n)}$ we denote the conjugates over K of $a \in L$) is non-zero. Let N be a given positive integer. With every element

$$\boldsymbol{\xi} = [x_1, \ldots, x_N] \in L^N,$$

where

$$x_i = \sum_{j=1}^n x_{ij}\omega_j \quad (x_{ij} \in K, i = 1, 2, \ldots, N)$$

we associate

$$\hat{\boldsymbol{\xi}} = [x_{11}, \ldots, x_{1n}, x_{21}, \ldots, x_{Nn}] \in K^{Nn}.$$

Note that the mapping $\Psi : \boldsymbol{\xi} \mapsto \hat{\boldsymbol{\xi}}$ of L^N in K^{nN} is one-to-one.

For every polynomial $F \in L[X_1, \ldots, X_N]$ the equality

$$F(\sum_{j=1}^n X_{1j}\omega_j, \ldots, \sum_{j=1}^n X_{Nj}\omega_j) = \sum_{j=1}^n \hat{F}_j(X_{11}, X_{12}, dots, X_{nN})\omega_j$$

defines an n-tuple $\hat{F}$ of polynomials $[\hat{F}_1, \ldots, \hat{F}_n]$ in nN variables over K.

Let now

$$\Phi = < F_1(X_1, X_2, \ldots, X_N), \ldots, F_N(X_1, X_2, \ldots, X_N) > : L^N \longrightarrow L^N$$

be an admissible polynomial mapping with $F_j \in L[X_1, \ldots, X_N]$. In view of Theorem 9.1 we may assume that all polynomials F_i are homogeneous. Associating with every F_j the just defined n-tuple $\hat{F}_{j1}, \ldots, \hat{F}_{jn}$ we obtain a polynomial mapping $\hat{\Phi} : K^{nN} \longrightarrow K^{nN}$. Clearly the polynomials defining $\hat{\Phi}$ are all homogeneous and non-linear, hence we have only to check that they do not have a non-trivial common zero in a fixed algebraic closure M of the field L. This will follow from the following lemma:

Lemma 9.5. *If L/K is separable of degree n and $F_1, \ldots, F_N \in L[X_1, \ldots, X_N]$ are homogeneous polynomials without a non-trivial common zero in M then the polynomials $\hat{F}_{jk}$ $(j = 1, 2, \ldots, N; k = 1, 2, \ldots, n)$ do not have a non-trivial common zero in M either.*

Proof: There exist n embeddings $\sigma_1, \ldots, \sigma_n$ of L in M, which are equal to the identity map on K. Extend them to maps of $L[X_1, \ldots, X_N]$ into $M[X_1, \ldots, X_N]$ applying σ_i to each of the coefficients, and denote the resulting map again by σ_i. Assume that the polynomials $\hat{F}_{jk}$ $(j = 1, 2 \ldots, N; k = 1, 2, \ldots, n)$ have a non-trivial common zero $(\alpha_{11}, \ldots, \alpha_{nN})$ in M. Since

$$F_j(\sum_{i=1}^n X_{1i}\omega_i, \ldots, \sum_{i=1}^n X_{Ni}\omega_i) = \sum_{k=1}^n \dot{F}_{jk}(X_{11}, \ldots, X_{nN})\omega_k,$$

implies

$$\sigma_r(F_j)(\sum_{i=1}^n X_{1i}\sigma_r(\omega_i), \ldots, \sum_{i=1}^n X_{Ni}\sigma_r(\omega_i)) = \sum_{k=1}^n \dot{F}_{jk}(X_{11}, \ldots, X_{nN})\sigma_r(\omega_k)$$

for $r = 1, 2, \ldots, n$ we obtain, by putting $X_{rs} = \alpha_{rs}$, the existence of a common zero

$$[\sum_{j=1}^n \alpha_{1j}\sigma_r(\omega_j), \ldots, \sum_{j=1}^n \alpha_{Nj}\sigma_r(\omega_j)]$$

of polynomials $\sigma_r(F_1), \ldots, \sigma_r(F_N)$.

Since the polynomials $F_1, \ldots, F_N$ have in M only the trivial common zero, the same applies to $\sigma_r(F_1), \ldots, \sigma_r(F_N)$ and thus we must have

$$\sum_{j=1}^n \alpha_{ij}\omega_j^{(r)} = \sum_{j=1}^n \alpha_{ij}\sigma_r(\omega_j) = 0$$

for $r = 1, 2, \ldots, n$ and $i = 1, 2, \ldots N$. For any fixed i we get a linear system with non-vanishing discriminant $\det[\omega_i^{(j)}]_{i,j=1,\ldots,n}$ and $\alpha_{ij} = 0$ for all i, j follows. $\square$

The theorem results immediately. $\square$

COROLLARY. *Every finite extension of the rationals has the properties* (**P**) *and* (**SP**).

PROOF: Apply theorems 9.3 and 9.4. $\square$

Note that in certain cases one can relax the condition of admissibility and still obtain the non-existence of fully invariant sets as shown in W.NARKIEWICZ ([64], II) and F.HALTER-KOCH,W.NARKIEWICZ [92b].

8. P.LIARDET [71] considered the following analogue of the property (**P**) for rational functions:

PROPERTY (**R**). *If f is a rational function of one variable with coefficients in K, and there exists an infinite subset A of K which is fully invariant, i.e. $f(X) = X$, then with suitable $\alpha, \beta, \gamma, \delta \in K$ one has*

$$f(T) = \frac{\alpha + \beta T}{\gamma + \delta T}.$$

It has been shown by W.NARKIEWICZ [63a] that the field of rational numbers has this property and P.LIARDET [71] showed that the same is true for all finitely generated fields. The analogues of problems XV, XVI and XVII for the property (**R**) are still open (**PROBLEM XVIII**).

Exercises

1. (W.NARKIEWICZ [62]) Let K be an algebraic number field and let $\omega_1, \ldots, \omega_n$ be an integral basis of K. For $\xi = (x_1 \omega_1 + \cdots + x_n \omega_n)/q$ (with $q, x_i \in Z$, $q > 0$ and $(q, x_1, \ldots, x_n) = 1$) put $f(\xi) = q$, $g(\xi) = \max \{|x_1|, \ldots, |x_n|\}$ and let $T \in K[X]$ be non-linear. Prove that if $X \subset K$ satisfies $T(X) = X$ then the conditions of Lemma 9.2 are satisfied and give thus another proof of property (**P**) for K.

2. (W.NARKIEWICZ [63a]) Let $X \subset \mathbf{Q}$ and let f be a rational function satisfying $f(X) = X$. Prove that if X is infinite, then there exist $a, b, c, d \in Z$ such that

$$f(X) = \frac{a + bX}{c + dX}.$$

3. Prove that p-adic number fields do not have the property (**P**).

X. Heights and transcendental extensions

1. The aim of this section is to show that the properties **(P)** and **(SP)** are preserved under arbitrary purely transcendental extensions. First we collect certain technical results, concerning heights in fields. More on this topic may be found in [DG], Chapter 3.

Let K be a field and let $\mathcal{M} = \mathcal{M}_K$ be a family of *absolute values* of K, i.e. mappings $v : K^* \longrightarrow \mathbf{R}$, satisfying

(a) $v(0) = 0$ *and* $v(x) > 0$ *for* $x \neq 0$,

(b) *For all* $v \in \mathcal{M}$ *and* $x, y \in K$ *one has* $v(x + y) \leq v(x) + v(y)$ *and* $v(xy) = v(x)v(y)$,

(c) *For almost all* $v \in \mathcal{M}$ *(i.e. for all except finitely many) one has*

$$(10.1) \qquad\qquad v(x + y) \leq \max\{v(x), v(y)\},$$

(d) *For every* $x \in K$ *there are only finitely many* $v \in \mathcal{M}$ *with* $v(x) \neq 1$.

Two absolute values of a field K are called *equivalent* if they define on K the same topology. We shall always assume that $\mathcal{M}_K$ consists of inequivalent absolute values. An absolute value v satisfying (10.1) is called a *non-archimedean absolute value*. The remaining absolute values are called *archimedean*. There are only finitely many archimedean absolute values, each equivalent to an absolute value defined by $v(x) = |\Phi(x)|$, where Φ is an embedding of K in $\mathbf{C}$. In sequel we shall consider only archimedean values so defined.

2. One says that $\mathcal{M}$ satisfies the *product formula* if there is a sequence λ_v of positive numbers such that for all non-zero $x \in K$ one has

$$\prod_{v \in \mathcal{M}} v(x)^{\lambda_v} = 1.$$

In that case one defines for every $v \in \mathcal{M}$ the *norm* $\| \cdot \|_v$ by putting

$$\| x \|_v = v(x)^{\lambda_v}$$

for all $x \in K$. One shows that the norms so defined satisfy the following conditions:

(A) $\| 0 \|_v = 0$ *and* $\| x \|_v > 0$ *for* $x \neq 0$,

(B) *For all $x, y \in K$ one has*

$$\| xy \|_v = \| x \|_v \| y \|_v ,$$

(C) *For archimedean $v \in M$ one has*

$$\lambda_v = \begin{cases} 1 & \text{if } v \text{ corresponds to an embedding of } K \text{ in } \mathbf{R}, \\ 2 & \text{otherwise} \end{cases}$$

and for all $x, y \in K$

$$\| x + y \|_v \le \| x \|_v + \| y \|_v$$

holds in the first case and

$$\| x + y \|_v \le 2(\| x \|_v + \| y \|_v)$$

in the second. In the last case one has, more generally,

$$\| x_1 + x_2 + \cdots + x_n \|_v \le 2^{n-1}(\| x_1 \|_v + \| x_2 \|_v + \cdots + \| x_n \|_v)$$

for all $x_1, x_2, \ldots, x_n \in K$.

(D) *For non-archimedean $v \in M$ one has*

$$\| x + y \|_v \le \max\{\| x \|_v, \| y \|_v\},$$

(E) *For every $x \in K$ there are only finitely many $v \in M$ with $\| x \|_v \ne 1$,*

(F) *For every non-zero $x \in K$ one has*

$$\prod_{v \in M} \| x \|_v = 1.$$

Note that for non-archimedean v the norm $\| \cdot \|_v$ is again an absolute value.

LEMMA 10.1. *Let $F(X_1, \ldots, X_n)$ be a polynomial of degree D with coefficients in K and let $\boldsymbol{\xi}_F = [\xi_1^{(F)}, \ldots]$ be the vector formed by all non-zero coefficients of F in a suitable K-space. If $v \in M_K$, then for all $x_i \in K$ one has*

$$\| F(x_1, \ldots, x_n) \|_v \le c_v \cdot \max_i \| \xi_i^{(F)} \|_v \max(1, \max_i \| x_i \|_v^D),$$

where

$$c = c_v(F) = \begin{cases} N2^{N-1} & \text{if } v \text{ is archimedean,} \\ 1 & \text{otherwise} \end{cases}$$

and N is the number of non-zero terms in F.

(Note that $N \le (D+1)^n$).

PROOF: Write

$$F(X_1, \ldots, X_n) = \sum_{i_1, \ldots, i_n} a_{i_1, \ldots, i_n} X_1^{i_1} \cdots X_n^{i_n}$$

with $a_{i_1, \ldots, i_n} \in K$. In the non-archimedean case the assertion follows immediately from (B) and (D) whereas for archimedean v one uses (B) and (C) to get

$$\| F(X_1,\ldots,X_n) \|_v \leq 2^{N-1} \sum_{i_1,\ldots,i_n} \| a_{i_1,\ldots,i_n} X_1^{i_1} \cdots X_n^{i_n} \|_v$$

$$\leq N 2^{N-1} \max_i \| \xi_i^{(F)} \|_v \max\{1, \max_i \| x_i \|_v^D\}.$$

$\square$

3. In the field $\mathbf{Q}$ of rational numbers we have the usual absolute value

$$v_\infty(r) = |r| \quad (r \in \mathbf{Q})$$

and for every prime p there is an absolute value v_p defined by $v_p(r) = \dfrac{1}{p^\alpha}$ where $\alpha = \alpha(r) \in \mathbf{Z}$ is defined for non-zero $r \in \mathbf{Q}$ by $r = p^\alpha \dfrac{a}{b}$, with $a, b \in \mathbf{Z}$ and $p \nmid ab$. One sees easily that if we put $\| r \|_v = v(r)$, then the product formula holds with $\mathcal{M}_\mathbf{Q} = \{v_\infty, v_2, v_3, \ldots, v_p, \ldots\}$ and $\lambda_v = 1$ for all v.

Now let $k = k_0(T)$ be the field of rational functions in one variable over an arbitrary field k_0. Here we shall consider the set $\mathcal{M}_k$ consisting of absolute values defined in the following way:

For any monic irreducible polynomial $P \in k_0[T]$ and non-zero $f \in k$ one puts

$$v_P(f) = e^{-\alpha \deg(P)},$$

where $\alpha \in \mathbf{Z}$ is defined by $f(T) = aP^\alpha g(T)/h(T)$, with $a \in k$, $g, h \in k_0[T]$ and $(gh, P) = 1$.

Moreover, one defines v_∞ by putting for every $f = g/h \in K$, with $f, g \in k_0[T]$

$$v_\infty(f) = e^{\deg g - \deg h}.$$

One sees immediately, that the set of all absolute values just defined satisfies the product formula with $\lambda_v = 1$ and $\| x \|_v = v(x)$.

4. If K/k is a finite separable extension then one can extend every absolute value v of k to an absolute value of K by means of the following result of E.ARTIN, G.WHAPLES [45]:

LEMMA 10.2. *Let M_k be a set of absolute values of a field k for which the product formula holds with suitable exponents λ_v. Let K/k be a finite separable extension of degree N and denote by $\mathcal{M}_K$ the set of all absolute values of K which prolong the absolute values of $\mathcal{M}_k$. Then the following is true:*

(i) *For every $v \in \mathcal{M}_k$ the set of all $w \in \mathcal{M}_K$ prolonging v is finite and if N_v denotes its cardinality, then*

$$1 \leq N_v \leq [K : k],$$

(ii) *The product formula holds for $\mathcal{M}_K$ with a suitable choice of exponents λ_w,*

(iii) If $v \in \mathcal{M}_k$ and $w_1, \ldots, w_r \in \mathcal{M}_K$ are all prolongations of v to K then

$$\sum_{i=1}^{r} \lambda_{w_i} = [K : k]\lambda_v.$$

For the proof see e.g. E.ARTIN, G.WHAPLES [45] or [H] § 20. $\square$

Recall for later use that in the case when K is a finite extension of the rationals every non-archimedean absolute value v corresponds to a prime ideal P_v of the ring $\mathbf{Z}_K$ of integers of K, absolute values prolonging the same absolute value of $\mathbf{Q}$ correspond to prime ideals dividing the same rational prime number p and the exponents λ_v are given by $\lambda_v = e_v f_v$, where $N(P_v) = p^{f_v}$ and $P_v^{e_v} \parallel p\mathbf{Z}_K$.

5. If K is a field with the product formula for a set $\mathcal{M}_K$ of absolute values then for $n = 1, 2, \ldots$ one defines the *height* $H_K(\mathbf{x})$ of a point $\mathbf{x} = [x_1, \ldots, x_n] \in K^n$ by

$$(10.2) \qquad H_K(\mathbf{x}) = \prod_{v \in \mathcal{M}_K} \max\{1, \parallel x_1 \parallel_v, \ldots, \parallel x_n \parallel_v\}.$$

LEMMA 10.3. *Let k be a field with product formula, let K/k be finite and separable and let L be a field satisfying $k \subset K \subset L$. Then for all $\mathbf{x} \in K^n$ one has*

$$H_L(\mathbf{x})^{1/[L:k]} = H_K(\mathbf{x})^{1/[K:k]}.$$

PROOF: For any fixed $v \in \mathcal{M}_K$ let $w_1, \ldots, w_r$ be all prolongations of v to L, denote by λ_{w_i} the corresponding exponents in the product formula for L and let λ be the exponent corresponding to v in the product formula for K. Then for any $\mathbf{x} = [x_1, \ldots, x_n] \in K^n$ we get in view of Lemma 10.2 (iii)

$$\prod_{i=1}^{r} \max\{1, \parallel x_1 \parallel_{w_i}, \ldots, \parallel x_n \parallel_{w_i}\} = \prod_{i=1}^{r} \max\{1, w_i(x_1)^{\lambda_{w_i}}, \ldots, w_i(x_n)^{\lambda_{w_i}}\}$$

$$= \prod_{i=1}^{r} \max\{1, v(x_1)^{\lambda_{w_i}}, \ldots, v(x_n)^{\lambda_{w_i}}\} = \prod_{i=1}^{r} \max\{1, v(x_1), \ldots, v(x_n)\}^{\lambda_{w_i}}$$

$$= \max\{1, v(x_1), \ldots, v(x_n)\}^{[L:K]\lambda} = \max\{1, \parallel x_1 \parallel_v, \ldots, \parallel x_n \parallel_v\}^{[L:K]}$$

and multiplying over all v we arrive at

$$H_L(\mathbf{x}) = H_K(\mathbf{x})^{[L:K]}$$

and thus

$$H_L(\mathbf{x})^{1/[L:k]} = H_K(\mathbf{x})^{[L:K]/[L:k]} = H_K(\mathbf{x})^{1/[K:k]}$$

as asserted. $\square$

COROLLARY. *If k is as in the lemma, $\dot{k}$ is its fixed separable closure and for $n \geq 1$ and $\mathbf{x} = [x_1, x_2, \ldots, x_n] \in \dot{k}^n$ we put*

$$h(\mathbf{x}) = H_K(\mathbf{x})^{1/[K:k]},$$

where K/k is a finite separable extension containing $x_1, \ldots, x_n$, then the value $h(\mathbf{x})$ does not depend on K and hence the mapping

$$h : \bigcup_{n=1}^{\infty} \hat{k}^n \longrightarrow \mathbf{R}$$

is well-defined.

PROOF: Immediate. $\square$

We shall call h the *absolute height*. In the next theorem we collect some properties of the heights H_K and h:

THEOREM 10.4. (i) If $k = \mathbf{Q}$, ξ is an algebraic number of degree N over $\mathbf{Q}$,

$$F(X) = a_N X^N + a_{N-1} X^{N-1} + \cdots + a_0 \quad (a_i \in \mathbf{Z}, a_N > 0)$$

is its minimal polynomial (with $(a_0, a_1, \ldots, a_N) = 1$) and $\xi_1 = \xi, \xi_2, \ldots, \xi_n$ are all roots of F then

$$h(\xi) = (a_N \prod_{j=1}^{N} \max(1, |\xi_j|))^{1/N}.$$

(ii) If $\hat{\mathbf{Q}}$ denotes the field of all algebraic numbers and for every $\mathbf{x} = [x_1, x_2, \ldots, x_n]$ in $\hat{\mathbf{Q}}^n$ we put $\Delta(\mathbf{x}) = [\mathbf{Q}(x_1, \ldots, x_n) : \mathbf{Q}]$ then for every positive C the set

$$\{\mathbf{x} \in \hat{\mathbf{Q}}^n : h(\mathbf{x}) + \Delta(\mathbf{x}) \leq C\}$$

is finite.

(iii) If $K = k(X)$ is the field of rational functions in one variable over a field k and $\delta(\boldsymbol{\xi})$ is defined for $\boldsymbol{\xi} = [f_1/q, f_2/q, \ldots, f_n/q] \in K^n$ (with $f_1, \ldots, f_n, q$ being polynomials over k satisfying $(f_1, \ldots, f_n, q) = 1$) by

$$\delta(\boldsymbol{\xi}) = \max\{\deg f_1, \ldots, \deg f_n, \deg q\},$$

then

$$H_K(\boldsymbol{\xi}) = e^{\delta(\boldsymbol{\xi})}.$$

PROOF: (i) Let K be a normal extension of $\mathbf{Q}$ containing $\mathbf{Q}(\xi)$, and put $M = [K : \mathbf{Q}(\xi)]$. Then $[K : \mathbf{Q}] = MN$ and thus the assertion can be written in the form

$$H_K(\xi) = (a_N \prod_{j=1}^{N} \max(1, |\xi_j|))^M.$$

In view of the equality

$$\prod_{\substack{v \\ \text{arch.}}} \max(1, \| \xi \|_v) = (\prod_{j=1}^{N} \max(1, |\xi_j|)^M$$

resulting from the fact that every ξ_j is a fixpoint of exactly M elements of the

Galois group of $K/\mathbf{Q}$, the assertion is equivalent to the equality

$$\text{(10.3)} \qquad \prod_{\substack{v \\ \text{nonarch.}}} \max(1, \| \xi \|_v) = a_N.$$

The proof of (10.3) utilizes a form of *Gauss's Lemma*. To state it extend an absolute value v of a field K to $K[X]$ by putting for $f(X) = \sum_{j=0}^{N} a_j X^j \in K[X]$, $v(f) = \max_j v(a_j)$.

LEMMA 10.5. *If v is nonarchimedean then for all polynomials $f_1, f_2 \in K[X]$ one has $v(f_1 f_2) = v(f_1) v(f_2)$.*

PROOF: Write $f_1(X) = \sum_{i=0}^{r} \alpha_i X^i$ and $f_2(X) = \sum_{i=0}^{s} \beta_i X^i$ and let $v(f_1) = v(\alpha_m)$ and $v(f_2) = \beta_n$. We may assume that one has

$$\text{(10.4)} \qquad v(\alpha_m) > v(\alpha_i), \quad v(\beta_n) > v(\beta_j)$$

for $i = m+1, m+2, \ldots, r$ and $j = n+1, n+2, \ldots, s$. Then $f_1 f_2(X) = \sum_{k=0}^{r+s} \gamma_k$ with $\gamma_k = \sum_{i+j=k} \alpha_i \beta_j$, thus $v(\gamma_k) \leq v(\alpha_m) v(\beta_n)$ results. On the other hand, since from $i + j = m + n$ and $[i, j] \neq [m, n]$ follows that either $i > m$ or $j > n$ must hold, thus (10.4) implies

$$v(\gamma_{m+n}) = v\Big(\alpha_m \beta_n + \sum_{\substack{i+j=m+n \\ i>m}} \alpha_i \beta_j + \sum_{\substack{i+j=m+n \\ j>n}} \alpha_i \beta_j\Big)$$

$$= v(\alpha_m \beta_n) = v(\alpha_m) v(\beta_n) = v(f_1) v(f_2). \quad \square$$

If we write the fractional ideal generated by ξ in the form I/J, where I, J are relatively prime ideals in the ring of integers of K and v is non-archimedean, then $\| \xi \|_v > 1$ holds if and only if J is divisible by the prime ideal P_v inducing v. If $P_v^c \| J$, then the remarks at the end of subsection 4 give $\| \xi \|_v = N(P_v)^c$ and this leads to

$$\text{(10.5)} \qquad \prod_{\substack{v \\ \text{nonarch.}}} \max(1, \| \xi \|_v) = N(J).$$

Applying the lemma to the factorization

$$F(X) = a_N \prod_{j=0}^{N} (X - \xi_j)$$

and to the norm $\| \cdot \|_v$ corresponding to a non-archimedean v we obtain

$$1 = \| F \|_v = \| a_N \|_v \prod_{j=1}^{N} \max(1, \| \xi_j \|_v),$$

because $(a_0, a_1, \ldots, a_N) = 1$ implies $\| F \|_v = 1$ for every non-archimedean v.

Multiplying over all such v's we get

$$1 = \prod_{\substack{v \\ \text{nonarch.}}} \| a_N \|_v \prod_{j=1}^{N} \prod_{\substack{v \\ \text{nonarch.}}} \max(1, \| \xi_j \|_v),$$

and in view of

$$\prod_{\substack{v \\ \text{nonarch.}}} \| a_N \|_v = \frac{1}{a_N^{MN}}$$

we arrive at

(10.6)
$$\prod_{j=1}^{N} \prod_{\substack{v \\ \text{nonarch.}}} \max(1, \| \xi_j \|_v) = a_N^{MN}.$$

If we denote by $J_1 = J, J_2, \ldots, J_M$ the ideals of $\mathbf{Z}_K$ conjugated to J then (10.5) implies

$$\prod_{\substack{v \\ \text{nonarch.}}} \prod_{j=1}^{N} \max(1, \| \xi_j \|_v) = \prod_{j=1}^{N} \prod_{\substack{v \text{ nonarch.} \\ \| \xi_j \|_v > 1}} \| \xi_j \|_v$$

$$= \prod_{j=1}^{N} N(J_j) = N(\prod_{j=1}^{N} J_j) = N(N_{K/\mathbf{Q}}(J)) = N(J)^{MN},$$

where $N_{K/\mathbf{Q}}$ denotes the ideal norm from K to $\mathbf{Q}$.

This equality jointly with (10.6) gives now $N(J) = a_N$ and the assertion results now from (10.3) and (10.5).

To prove (ii) note first that if $h(\mathbf{x}) + \Delta(\mathbf{x}) \leq C$ and $K = Q(x_1, \ldots, x_n)$ then $H_K(\mathbf{x})$ is bounded and in view of $h(\mathbf{x}) \geq h(x_i)$ for all i we get from (i) and Vieta's formulas the boundedness of the coefficients of the minimal polynomial (over $\mathbf{Z}$) of x_i for $i = 1, 2, \ldots, n$. Thus there are only finitely many possibilities for $\mathbf{x}$.

The assertion (iii) is a direct consequence of the definition of H_K and the product formula. $\square$

6. Our main tool will be the following lemma which in this form is due to P.LIARDET [71] and appeared for the first time with another definition of height in a special case ($k = \mathbf{Q}$ and all polynomials of the same degree) in D.G.NORTHCOTT [49a],[49b],[50].

LEMMA 10.6. *Let K be a field with product formula and let $\hat{K}$ be its separable closure. Let $\Phi = [F_1, F_2, \ldots, F_n] : \hat{K}^n \longrightarrow \hat{K}^n$ be a polynomial mapping defined by homogeneous polynomials F_i of n variables with coefficients in K and denote by d, D the minimal resp. maximal degree of the F_i's.*

(i) *There exists a constant $C = C(\Phi)$ such that for all $x \in \hat{K}^n$ one has*

$$h(\Phi(\mathbf{x})) \leq Ch(\mathbf{x})^D.$$

(ii) *If the polynomials $F_1, \ldots, F_n$ do not have a non-trivial common zero in K, then there exists a positive constant $c = c(\Phi)$ such that*

$$h(\Phi(\mathbf{x})) \geq c h(\mathbf{x})^d$$

holds for all $\mathbf{x} \in \hat{K}^n$.

PROOF: The assertion (i) is a direct consequence of Lemma 10.1. To prove (ii) we apply Hilbert's Nullstellensatz which gives for every $i = 1, 2, \ldots, n$ the existence of polynomials $A_j^{(i)} \in K[X_1, \ldots, X_n]$ $(j = 1, 2, \ldots, n)$ and a positive integer μ such that

$$(10.7) \qquad \sum_{j=1}^{n} A_j^{(i)}(X_1, \ldots, X_n) F_j(X_1, \ldots, X_n) = X_i^{\mu}.$$

Let $[x_1, \ldots, x_n] \in \hat{k}^n$ be given, put $L = K(x_1, \ldots, x_n)$ and let $v \in \mathcal{M}_L$. If

$$\max_j (\| x_j \|_v \leq 1$$

then

$$(10.8) \qquad \begin{aligned} &\max(1, \| F_1(x_1, \ldots, x_n) \|_v, \ldots, \| F_n(x_1, \ldots, x_n) \|_v) \\ &\geq B_v \max(1, \| x_1 \|_v, \ldots, \| x_n \|_v) \end{aligned}$$

holds trivially with $B_v = 1$.

If $\max_j (\| x_j \|_v = \alpha > 1$ then choose $t \in L$ with $\| t \|_v = \dfrac{1}{\alpha} < 1$ and put $y_i = t x_i$ for $i = 1, 2, \ldots, n$, so that $\max_i(\| y_i \|_v) = 1$. The equality (10.7) and Lemma 10.1 give now

$$1 = \max_i(\| y_i \|_v^{\mu}) \leq c_v \max(1, \| y_1 \|_v^A, \ldots, \| y_n \|_v^A) \cdot \max_j(\| F_j(y_1, \ldots, y_n) \|_v)$$

$$= c_v \max_j(\| F_j(y_1, \ldots, y_n) \|_v),$$

where A denotes the maximum of the degrees of the polynomials A_i^j and c_v is a constant, which for almost all v equals 1. Since the polynomials F_j are homogeneous we get

$$1 \leq c_v \cdot \max_j(\| t \|_v^{\deg F_j} F_j(x_1, \ldots, x_n)) \leq c_v \| t \|_v^d \max_j(\| F_j(x_1, \ldots, x_n) \|_v)$$

and finally

$$\max_j(\| F_j(x_1, \ldots, x_n) \|_v) \geq c_v^{-1} \max_j(\| x_j \|_v^d)$$

thus (10.8) holds with $B_v = c_v^{-1}$. Since for almost all v's we have $B_v = 1$, mutiplying the inequalities (10.8) we arrive at the assertion. $\square$

7. As an application we shall now describe a property stronger than (**P**), which has been introduced by P.LIARDET [70], and present his proof that all algebraic number fields enjoy it. This result permitted the construction of algebraic extensions of the rationals of infinite degree which have the property (**P**).

A field K is said to have the *property* $(\bar{\mathbf{P}})$ if for every non-linear polynomial $f \in K[X]$ and every subset X of the algebraic closure $\bar{K}$ of K consisting of elements of bounded degree over K, the equality $f(X) = X$ implies the finiteness of X.

THEOREM 10.7. (P.LIARDET [70]) (i) *The property $(\bar{\mathbf{P}})$ is preserved by arbitrary finite extensions.*

(ii) *Every finite extension of the rationals has the property $(\bar{\mathbf{P}})$.*

PROOF: (i) Follows immediately from the definition of $(\bar{\mathbf{P}})$.

(ii) In view of (i) it suffices to consider the case of the rational number field **Q**. In this case the assertion follows from the Corollary to Lemma 9.2 in which we put $f(x) = H_{\mathbf{Q}}(x)$ and use Theorem 10.4 (ii) and Lemma 10.6 (ii) to check that its assumptions are satisfied. □

In the quoted paper it has been also shown that the property $(\bar{\mathbf{P}})$ is preserved by arbitrary purely transcendental extensions and thus every finitely generated field has the it. The question, whether the property $(\bar{\mathbf{P}})$ is equivalent to one of the properties considered in Section IX remains unanswered (**PROBLEM XIX**).

8. Theorem 10.7 can be applied to produce an example of a rather large algebraic extension of **Q** with the property (**P**). This example, due to K.K.KUBOTA, P.LIARDET [76], refuted a conjecture of W.NARKIEWICZ [63b].

THEOREM 10.8. *There exists an algebraic extension of* **Q** *having the property* (**P**) *which cannot be generated over the rationals by a set of elements of bounded degree.*

PROOF: Let $F_1, F_2, \ldots$ be the sequence of all non-linear polynomials with algebraic coefficients and let $p_1 < p_2 < \ldots$ be the sequence of all rational primes.

Denote by E_n the set of all algebraic numbers which are cyclic elements for some polynomial F_i with $i \leq n$.

LEMMA 10.9. *For any integer D and any n the set of all elements $x \in E_n$ with $\deg x \leq D$ is finite.*

PROOF: Each such set is the union of finitely many sets

$$\{x : x \in Cycl(F_j), \deg x \leq D\}$$

which are finite by Theorem 10.7. □

Now we construct an ascending sequence of finite extensions of the rationals: put $K_0 = \mathbf{Q}$ and if K_{n-1} is already defined then choose K_n to be an extension of K_{n-1} of degree p_n which satisfies

$$K_n \cap E_n = K_{n-1} \cap E_n.$$

Such choice is always possible, since every algebraic number field L has always an extension of a given degree which does not contain a given finite set of algebraic numbers disjoint with L. In our case the set to be avoided consists of elements $a \in E_n$ having their degree over K_{n-1} not exceeding p, thus $\deg_Q a \leq p[K_{n-1} : Q]$ and the finiteness of this set is implied by the preceding lemma.

Finally we put

$$K = \bigcup_{n=1}^{\infty} K_n.$$

The extension K/Q is algebraic by construction and cannot be generated by a set of algebraic numbers of bounded degree, since it contains elements of degree divisible by p for any given prime number p.

To show that K has the property (P) let $f = F_s \in K[X]$ be a non-linear polynomial of degree $d > 1$ and assume that the set $A \subset K$ is fully invariant with respect to f. Denote by r the smallest integer with $f \in K_r[X]$. Let $x \in A$ and choose n so large that the conditions $x \in K_n$ and $p_n > d$ will be satisfied. Construct a sequence $\{x_m\}$ by putting $x_0 = x$, $x_{m+1} = f(x_m)$ for $m = 0, 1, \ldots$ and defining inductively $x_{-1}, x_{-2}, \ldots$ lying in A by the same formula. The degree N of the extension $K_m(x_{-j-1})/K_m(x_{-j})$ is for $j = 0, 1, \ldots$ a divisor of d, thus for $n = 1, 2, \ldots$ the degree $D = [K_m(x_{-n}) : K_m]$ cannot be divisible by a prime $\geq p_m$. But $x_{-n} \in X \subset K$ thus for a suitable $t \geq m$ we have $x_{-n} \in K_t$, however in view of $[K_t : K_m] = p_{m+1} \cdots p_t$ we obtain that $D = [K_t : K_m]/[K_t : K_m(x_{-n})]$ is a divisor of $p_{m+1} \cdots p_t$. Thus we must have $x_{-n} \in K_m$ for $n = 0, 1, \ldots$ and since by Corollary to Theorem 9.4 K_m has the property (P), $x + 0$ lies in a finite cycle. This implies, by the construction of K

$$X \subset Cycl(f) \cap K = Cycl(F_s) \cap K = Cycl(F_s) \cap K_s \subset K_s.$$

If we put $M = r + s$ then $X \subset K_M$ and $f \in K_M[X]$ hence the application of Corollary to Theorem 9.4 leads to the finiteness of A. $\square$

9. Now we show that the properties (P) and (SP) are preserved under arbitrary purely transcendental extensions, i.e. extensions K/k with $K = k(\{X_\alpha\})$ where $\{X_\alpha\}$ is a set of elements algebraically independent over k, and begin with the property (P).

THEOREM 10.10. (W.NARKIEWICZ [62],I) Let k be a field having the property (P) and let $K = k(\mathcal{X})$, where $\mathcal{X} = \{X_\alpha\}$ is a set of elements algebraically independent over k of arbitrary cardinality. Then K has also the property (P).

PROOF: Our first aim will be the reduction of the proof to the case when $\mathcal{X}$ consists of a single element and for this we need a simple lemma of E.STEINITZ [10]:

LEMMA 10.11. Let $K = k(X)$ be the field of rational functions in one variable X over a field k. For $r = P/Q \in K$, where P, Q are relatively prime polynomials over k, we put (as in Theorem 10.4 (iii))

$$\delta(r) = \max(\deg P, \deg Q).$$

If f is a polynomial in one variable with coefficients in k and N is its degree, then for every $r \in K$ one has

$$\delta(f(r(X))) = N\delta(r(X)).$$

PROOF: Write $f(u) = \sum_{j=0}^{N} a_j u^j$ $(a_j \in k)$. If $r(X) = P(X)/Q(X) \in K$ with $(P, Q) = 1$, then

$$f(r) = \frac{\sum_{j=0}^{N} a_j P^j Q^{N-j}}{Q^N}$$

and if $\Delta = (\sum_{j=0}^{N} a_j P^j Q^{N-j}, Q^N)$ and π is an irreducible factor of Δ, then π divides both Q and $a_N P^N$, thus must be constant and $\Delta = 1$ results. Hence

$$\delta(f(r(X))) = \max(\deg Q^N(X), \deg (\sum_{j=0}^{N} a_j P^j(X) Q^{N-j}(X))),$$

and the assertion becomes evident. $\square$

LEMMA 10.12. (W.NARKIEWICZ [62],II) *If the assertion of the theorem holds for the case* $\mathcal{X} = \{X\}$, *then it holds also for arbitrary* $\mathcal{X}$.

PROOF: If (**P**) is preserved under a simple transcendental extension, then it is preserved also under every purely transcendental extension of finite rank. Let now $\mathcal{X} = \{X_\alpha\}$ be an infinite set algebraically independent over k and let $K = k(\mathcal{X})$. Assume that there is an infinite subset $A \subset K$ and a non-linear polynomial $f \in K[t]$ of degree $N > 1$ such that $f(A) = A$. We may safely assume $f \in k[t]$, since otherwise we might adjoin to k all indeterminates occuring in the coefficients of f, without affecting the property (**P**). Let ξ be an element of A which does not lie in k, let X_1 be an indeterminate on which ξ depends, put $M = k(\{X_\alpha : \alpha \neq 1\})$ and let $\delta : M(X_1) \longrightarrow \mathbf{Z}$ be the function defined in Lemma 10.11. Assume that

$$\delta(\xi) = \min\{\delta(\Xi) : \Xi \in A \setminus M\}.$$

Note that $\xi \notin M$ and thus $\delta(\xi) \geq 1$. There exists $\eta \in A$ satisfying $f(\eta) = \xi$ and Lemma 10.11 implies

$$\delta(\eta) = \frac{\delta(\xi)}{N} < \delta(\xi)$$

which by our choice of ξ is possible only if η lies in M. But then $\xi = f(\eta) \in M$ follows, giving a contradiction. $\square$

10. It remains to establish the assertion of the theorem for simple transcendental extensions $K = k(X)$. Let thus $A \subset K$ and assume that with a certain $F \in K[T]$ of degree $N \geq 2$ one has $F(A) = A$. Our tool will be the Corollary to Lemma 9.2 in which we put $T = F$ and take for f the function δ defined above. The truth of the second condition of that Corollary follows from Lemma 10.6 (ii) and so it remains to check the first condition. It holds obviously for finite k, so we can assume k to be infinite. Write

$$F(T) = \frac{\sum_{j=0}^{N} a_j(X)T^j}{\Delta(X)},$$

with $\Delta, a_0, \ldots, a_N \in K[X]$, $a_N \neq 0$, and choose an infinite sequence r_i of distinct elements of k, satisfying

$$\Delta(r_i)a_N(r_i) \neq 0 \quad (i = 1, 2, \ldots).$$

For every i put

$$A_i = \left\{ \frac{P(r_i)}{Q(r_i)} \; : \; \frac{P}{Q} \in A, Q(r_i) \neq 0 \right\} \subset k$$

and

$$F_i(T) = \frac{1}{\Delta(r_i)} \sum_{j=0}^{N} a_j(r_i)T^j \in k[T].$$

Each polynomial F_i is of degree N and one sees easily that $F_i(A_i) = A_i$ holds. As $A_i \subset k$ and k has the property (**P**), the finiteness of each of the sets A_i results. Since there can be only finitely many rational functions with bounded degrees of numerator and denominator, which at each point of an infinite sequence attain values from a finite set, the conditions of the Corollary to Lemma 9.2 are satisfied and the application of that lemma concludes the proof of the theorem. $\square$

11. We turn now to the property (**SP**).

THEOREM 10.13. (F.HALTER-KOCH,W.NARKIEWICZ [92a]) *The property* (**SP**) *is preserved under arbitrary purely transcendental extensions.*

PROOF: We deal first with finitely generated extensions and start with the simplest case:

LEMMA 10.14. *If a field k has the property* (**SP**), *so does $k(T)$, where T is transcendental over k.*

PROOF: Let $K = k(T)$ and let Φ be an admissible polynomial mapping $K^n \longrightarrow K^n$ defined by the polynomials $f_1, \ldots, f_n$. In view of Theorem 9.1 we may assume that all polynomials f_i are homogeneous. Assume that for a set $A \subset K^n$ one has $\Phi(A) = A$. Lemma 10.6.(ii) implies the existence of a constant $C > 0$ such that if $\boldsymbol{\xi} \in K^n$ and $H_K(\boldsymbol{\xi}) \geq C$ then $H_K(\Phi(\boldsymbol{\xi})) > H_K(\boldsymbol{\xi})$, where $H_K(\boldsymbol{\xi}) = e^{\delta(\boldsymbol{\xi})}$ with $\delta(\boldsymbol{\xi})$ defined as in Theorem 10.4 (iii).

Put

$$A_0 = \{\boldsymbol{\xi} \in A : \delta(\boldsymbol{\xi}) < C\}.$$

To obtain the finiteness of A it suffices now, in view of Lemma 9.2 (b), to show that A_0 is finite. By the Nullstellensatz of Hilbert we obtain the existence of polynomials $A_j^{(k)}$ over K and of an exponent ρ satisfying

$$(10.9) \qquad \sum_{j=1}^{n} f_j(X_1,\ldots,X_n)A_j^{(k)}(X_1,\ldots,X_n) = X_k^{\rho} \quad (k=1,\ldots,n).$$

Let D be the set of all non-zero coefficients (which are rational functions in T over k) of the polynomials f_j, $A_j^{(k)}$ and put $d = \max\{\delta(u) : u \in D\}$. Observe now that the field k is neither algebraically closed nor real closed, since in that case it could not have the property (**SP**). Hence there exist algebraic extensions of k of arbitrarily large degrees and in particular there are infinitely many elements α of $\hat{k}$, the algebraic closure of k, with $\deg_k \alpha > d + 2C$. This choice guarantees that none of the rational functions in D have α for its pole or zero and in particular we are allowed to put $T = \alpha$ in the identity (10.9). This gives

$$(10.10) \qquad \sum_{j=1}^{n'} \hat{f}_j(X_1,\ldots,X_n)\hat{A}_j^{(k)}(X_1,\ldots,X_n) = X_k^{\rho} \quad (k=1,\ldots,n),$$

where $\hat{f}_j$ and $\hat{A}_j^{(k)}$ are polynomials over $k(\alpha)$. If we put

$$\hat{A}_0 = \{\boldsymbol{\xi}(\alpha) : \boldsymbol{\xi}(T) \in A_0\} \subset k(\alpha)$$

and

$$\hat{\Phi} = [\hat{f}_1,\ldots,\hat{f}_n],$$

then $\hat{\Phi}(\hat{A}_0) = \hat{A}_0$ and since due to $\deg \hat{f}_j = \deg f_j$ and (10.10) the map $\hat{\Phi}$ is admissible and by Theorem 9.4 the field $k(\alpha)$ has the property (**SP**), the set $\hat{A}_0$ must be finite. Since we have infinitely many possibilities for α the finiteness of A_0 results. It remains to invoke Lemma 9.2 (b) to arrive at the assertion of the lemma. $\square$

COROLLARY 1. *Every finitely generated purely transcendental extension of a field having the property (**SP**) has this property too.*

PROOF: Easy induction. $\square$

COROLLARY 2. *If k has the property (**SP**) and the extension K/k is finitely generated, then K has also the property (**SP**).*

PROOF: Follows from Theorem 9.4 and the preceding corollary. $\square$

COROLLARY 3. *If K is a field, finitely generated over its prime field, then it has the property (**SP**).*

PROOF: Finite fields have the property (**SP**) trivially and the field of the rationals has it due to Theorem 9.3. Now it suffices to apply the preceding Corollary. $\square$

12. Let $\mathcal{T} = \{T_\alpha\}$ be an infinite set algebraically independent over k and let $K = k(\mathcal{T})$. Let $\Phi = [f_1,\ldots,f_n] : K^n \longrightarrow K^n$ be an admissible polynomial map and assume that for a certain set $A \subset K^n$ one has $\Phi(A) = A$. We may assume in view of Theorem 9.1, that the f_i's are all homogeneous polynomials and moreover all their coefficients lie in k, since otherwise we could adjoin them to k and obtain, according to Corollary 2 to the last lemma, a field with the property (**SP**). We can also assume k to be infinite, since otherwise we could adjoin to k one of the T_α's.

Since Φ is admissible, Hilbert's Nullstellensatz gives the existence of a positive integer ρ and polynomials $A_j^{(k)}$ such that the equality (10.9) holds. Adjoining, if necessary, the coefficients of all $A_j^{(k)}$'s to k we may safely assume that they all lie in k. We need now two simple lemmas:

LEMMA 10.15. *If $K = k(\{X_\alpha\})$ is a purely transcendental extension and $\eta \in K$ is algebraic over k, then $\eta \in k$.*

PROOF: First let $K = k(X)$ with X transcendental over k and let

$$\eta(X) = \frac{P(X)}{Q(X)},$$

with $P, Q \in k[X]$, $(P, Q) = 1$. If $\eta(X)$ were algebraic over k, then with a certain $N \geq 1$ and $a_j \in k$, $a_N \neq 0$ we would have

$$0 = \sum_{j=0}^{N} a_j \frac{P^j(X)}{Q^j(X)},$$

thus

$$\sum_{j=0}^{N} a_j P^j(X) Q^{N-j}(X) = 0,$$

implying the divisibility of $a_n P(X)^n$ by $Q(X)$. Hence $Q(X) = c$ with some non-zero $c \in k$ thus

$$\sum_{j=0}^{N} a_j c^{N-j} P^j(X) = 0.$$

If $N \geq 1$ then looking at the leading coefficient of the left-hand side of the last equation we get $a_N = 0$, contrary to our assumption. Thus $N = 0$, so $P(X) = c_1 \in k$ and $\eta \in k$ results.

The case of a finitely generated purely transcendental extension follows now by recurrence and in the general case it suffices to observe that every element of $k(\mathcal{X})$ is contained in a finitely generated purely transcendental extension of k. □

LEMMA 10.16. *If $\boldsymbol{\xi} \in K^n$ satisfies $\Phi(\boldsymbol{\xi}) \in k^n$, then $\boldsymbol{\xi} \in k^n$.*

PROOF: Write $\boldsymbol{\xi} = [\xi_1, \ldots, \xi_n]$, $(\xi_i \in K)$ and observe that it suffices to establish the assertion in the case when $K = k(T)$ and then apply induction, since all ξ_i's lie in a finitely generated purely transcendental extension of k.

Write $\xi_i(T) = \dfrac{\alpha_i(T)}{\beta(T)}$ with $\alpha_i, \beta \in k(T)$ $(i = 1, 2, \ldots, n)$ and $(\alpha_1, \ldots, \alpha_n, \beta)$

$= 1$. Assume moreover that the polynomial β is monic. Let

$$\Phi(\xi) = c = [c_1, \ldots, c_n].$$

Then

(10.11) $\qquad f_i(\alpha_1, \ldots, \alpha_n) = \beta^{\deg f_i} f_i(\xi_1, \ldots, \xi_n) = \beta^{\deg f_i} c_i.$

If $\beta \neq 1$ and θ is a root of β in the algebraic closure $\hat{k}$ of k then (10.11) shows that $[\alpha_1(\theta), \ldots, \alpha_n(\theta)]$ is a common zero of $f_1, \ldots, f_n$, hence $\alpha_i(\theta) = 0$ $(i = 1, 2, \ldots, n)$ and this contradicts our assumptions. Thus $\beta = 1$ and $\xi_1(T), \ldots, \xi_n(T)$ are all polynomials. Consider the mapping Ψ induced by Φ in the corresponding projective space:

$$\Psi : P_{n-1}(\hat{k}) \longrightarrow P_{n-1}(\hat{k}).$$

Elementary algebraic geometry (see e.g. [SH], chapter I, §5, Theorem 8) shows that Ψ is a finite mapping, and this gives the finiteness of $\Psi^{-1}(c)$.

Our assumptions imply that $\Psi^{-1}(c)$ contains all elements $[\alpha_1(u), \ldots, \alpha_n(u)]$ with $u \in \hat{k}$. Thus there exists a finite set of, say, M elements:

$$\{[\theta_1^{(i)}, \ldots, \theta_n^{(i)}] \in \hat{k}^n : i = 1, 2, \ldots, M\}$$

such that for every $u \in \hat{k}$, with a suitable $\lambda = \lambda(u) \in k$ and $i = i(u) \leq M$ we have

$$\alpha_j(u) = \lambda \theta_j^{(i)} \quad (j = 1, 2, \ldots, n).$$

It follows that there exists an index $1 \leq i \leq M$ such that for infinitely many $u \in \hat{k}$ the equality

$$\alpha_j(u) = \lambda(u)\theta_j$$

holds for $j = 1, 2, \ldots, n$ with $\theta_j = \theta_j^{(i)}$. This implies that the polynomial $\gamma_j(T) = \dfrac{\alpha_j(T)}{\theta_j}$ does not depend on j and thus with a certain polynomial $\gamma(T)$ one has

$$\alpha_j(T) = \theta_j \gamma(T).$$

Finally we obtain

$$c_j = f_j(\alpha_1, \ldots, \alpha_n) = f_j(\theta_1 \gamma, \ldots, \theta_n \gamma) = \gamma^{\deg f_j} f_j(\theta_1, \ldots, \theta_n)$$

which shows that γ is algebraic over k and the same must be true for α_i $(i = 1, 2, \ldots, n)$. The application of Lemma 10.15 gives now $\alpha_i \in k$ and $\xi \in k^n$. $\square$

COROLLARY. *If for every finite subset* $T = \{T_{\alpha_1}, \ldots, T_{\alpha_r}\}$ *of* $\mathcal{T}$ *we put*

$$k_T = k(\{T_{\alpha_1}, \ldots, T_{\alpha_r}\})$$

then the set $A_T = A \cap k_T$ *is finite.*

PROOF: The lemma implies $\Phi(A_T) = A_T$ and the assertion follows from Corollary 1 to Lemma 10.14. $\square$

Now we can conclude the proof of Theorem 10.13. If $A \subset k^n$ then its finiteness follows from our assumptions.

Let $\boldsymbol{\xi}$ be an arbitrary element of $A \setminus k^n$, put $\Xi_0 = \boldsymbol{\xi}$ and for $i = 1, 2, \ldots$ let $\Xi_i = \Psi(\Xi_{i-1})$. The element $\boldsymbol{\xi}$ lies in A_T for a suitable finite $T \subset \mathcal{T}$, and our assumptions imply that all terms of the obtained sequence lie in A_T. The preceding corollary shows that the sequence Ξ_i can have only finitely many distinct terms and hence must contain a finite cycle Θ. If $\Theta \subset k^n$, then Lemma 10.10 would imply $\{\Xi_i\} \subset k^n$ and hence $\boldsymbol{\xi} \in k^n$, a contradiction. Since the coefficients of the polynomials defining Φ are assumed to be in k, for every k-automorphism σ of k_T of the form $T_{\alpha_i} \mapsto aT_{\alpha_i}$, with non-zero $a \in k$, the sequence $\sigma(\Theta)$ is a finite cycle in k_T and since k is infinite and Θ contains elements not belonging to k^n the union of all cycles so obtained form an infinite subset $\mathcal{Y}$ of k_T with $\Phi(\mathcal{Y}) = \mathcal{Y}$, contradicting Corollary 1 to Lemma 10.14. $\square$

Exercises

1. Prove that for non-zero $x \in k$ one has $h(1/x) = h(x)$.

2. (D.G.NORTHCOTT [49a]) Let $\mathbf{x} = [x_1, \ldots, x_n]$ with algebraic numbers x_i, put $K = \mathbf{Q}(x_1, \ldots, x_n)$ and define the *Northcott height* D_K by

$$D_K(\mathbf{x}) = \frac{\prod_\sigma (\sum_{j=1}^n |\sigma(x_j)|)}{N_{K/\mathbf{Q}}(I_\mathbf{x})},$$

where σ runs over all embeddings of K into the complex field and $I_\mathbf{x}$ denotes the fractional ideal in K generated by the components of $\mathbf{x}$.

(i) Show that if L is any algebraic number field containing all components of $\mathbf{x}$, then

$$D_K^{[K:\mathbf{Q}]}(\mathbf{x}) = D_L^{[L:\mathbf{Q}]}(\mathbf{x}),$$

hence

$$\Delta(\mathbf{x}) = D_K^{[K:\mathbf{Q}]}(\mathbf{x})$$

defines a map, which does not depend on K.

(ii) Prove that for any given n and C the set of all $\mathbf{x}$ satisfying

$$\Delta(\mathbf{x}) + [K : \mathbf{Q}] \leq C$$

is finite.

(iii) Prove the analogue of Lemma 10.6 in which h is replaced by Δ.

XI. Pairs of polynomial mappings

1. Theorem 9.3 implies that if A is an infinite subset of the rationals, $P \in \mathbf{Q}[X]$ and $P(A) = A$ then deg $P = 1$. The question has been posed whether two polynomials $P_1, P_2 \in \mathbf{Q}[X]$ acquiring the same set of values on an infinite subset of $\mathbf{Q}$ must be of the same degree (W.NARKIEWICZ [63b]). A negative answer has been provided by K.K.KUBOTA [72a]:

THEOREM 11.1. *Let $P_1 \in \mathbf{Q}[X]$ be a polynomial of degree $N \geq 1$ which is not injective on $\mathbf{Q}$, i.e. there exist two rationals $r < s$ with $P_1(r) = P_1(s)$. Then there exists a polynomial $P_2 \in \mathbf{Q}[X]$ of degree $2N$ and an infinite subset A of the rationals such that $P_1(A) = P_2(A)$ holds.*

PROOF: Let $F(X) = (X - r)^2 + X + s - r$ and put $P_2(X) = P_1(F(X))$. Then clearly deg $P_2 = 2 \deg P_1$. Put $a_0 = r$ and $a_{i+1} = F(a_i)$ for $i = 0, 1, \dots$. In particular $a_1 = s$. Since $F(X)$ increases in $[r, \infty)$, the set $A = \{a_0, a_1, \dots\}$ is infinite. It suffices now to observe that $F(A) \subset A$, thus $P_2(A) \subset P_1(A)$, and since for every $i \geq 1$ we have $P_1(a_i) = P_1(F(a_{i-1})) = P_2(a_{i-1}) \subset P_2(A)$ thus in view of $P_1(a_0) = P_1(r) = P_1(s) = P_1(a_1) = P_2(a_0) \subset P_2(A)$ we get $P_1(A) \subset P_2(A)$ and $P_1(A) = P_2(A)$ follows. $\square$

Note however that if $P_1, P_2 \in \mathbf{Q}[X]$ satisfy $P_1(\mathbf{Q}) = P_2(\mathbf{Q})$ then with suitable rational a, b one has $P_2(X) = P_1(aX + b)$, as shown in L.JANKOWSKI, A.MARLEWSKI [90]. The same assertion holds for polynomials $P_1, P_2 \in \mathbf{Z}[X]$ satisfying $P_1(\mathbf{Z}) = P_2(\mathbf{Z})$. This follows from a result of H.S.SHAPIRO [57], who proved that if $P_1(\mathbf{Z}) \subset P_2(\mathbf{Z})$ then there exists a polynomial R such that $P_1(X) = P_2(R(X))$.

2. The proof of the preceding theorem utilized the non-injectivity on the set A of the polynomial of lower degree. The next theorem, which is a particular case of a result of K.K.KUBOTA [72a], shows that without this condition Theorem 11.1 would cease to be true. Although it is a particular case of Theorem 11.5 we give a separate elementary proof of it.

THEOREM 11.2. *Let $P_1, P_2 \in \mathbf{Q}[X]$ with deg $P_1 < $ deg P_2 and assume that there is a subset A of the rationals such that $P_1(A) \subset P_2(A)$ and the restriction of P_1 to A is injective. Then A must be finite.*

PROOF: Observe first that it is sufficient to deal with polynomials P_1, P_2 having coefficients in $\mathbf{Z}$. In fact, if q denotes the common denominator of the coefficients of P_1, P_2, and for $i = 1, 2$ we put $F_i(X) = qP_i(X)$, then $F_i \in \mathbf{Z}[X]$ and $F_1(A) = F_2(A)$.

A sequence $x_0, x_1, \ldots, x_{r-1}$ of distinct rationals is called a *finite $P_1 - P_2$-cycle* provided for $i = 0, 1, \ldots, r - 1$ one has $P_2(x_{i+1}) = P_1(x_i)$ where we put additionally $x_r = x_0$. If $x_0, x_1, x_2, \ldots$ is an infinite sequence satisfying

(11.1) $$P_2(x_i) = P_1(x_{i-1})$$

then it is called an *infinite $P_1 - P_2$-cycle*.

LEMMA 11.3. *If the polynomial P_2 is monic, then every finite $P_1 - P_2$-cycle consists of rational integers and in every infinite $P_1 - P_2$-cycle at most finitely many terms are non-integral.*

PROOF: Write $P_1(X) = cX^n + \cdots$, $P_2(X) = x^m + \cdots$, and let $\{x_i\}$ be an $P_1 - P_2$-cycle. Write $x_i = p_i/q_i$ with coprime integers p_i, q_i and $q_i > 0$. Then

$$P_1(x_i) = \frac{cp_i^n + \cdots}{q_i^n} = P_2(x_{i+1}) = \frac{p_{i+1}^m + \cdots}{q_{i+1}^m},$$

and the last fraction is irreducible. This shows that with certain positive integers d_i we have

(11.2) $$q_i^n = d_i q_{i+1}^m.$$

If the cycle is finite, then by multiplying these equalities and denoting by D and Δ the products of the d_i's and q_i's respectively, we arrive at $\Delta^n = D\Delta^m$, hence $D\Delta^{m-n} = 1$ and thus D and all q_i's are equal to 1.

If the cycle is infinite, and for some i one has $q_{i+1} \geq q_i$ then $1 \geq d_i q_i^{m-n}$ follows and in view of $m > n$ we infer $q_i = 1$ and (11.2) gives now $1 = q_i = q_{i+1} = q_{i+2} = \ldots$. This shows also that $q_i > 1$ implies $q_{i+1} < q_i$ and we see that for sufficiently large i the number x_i is an integer. $\square$

LEMMA 11.4. *For arbitrary polynomials $P_1, P_2 \in \mathbf{Z}[X]$ with $n = \deg P_1 < m = \deg P_2$ the union of all finite $P_1 - P_2$-cycles is finite and there are no infinite $P_1 - P_2$-cycles.*

PROOF: We show first that it suffices to deal with the case when the polynomial P_2 is monic. Let a be the leading coefficient of P_2 and put

$$F_i(X) = a^{m-1} P_i(X/a) \quad (i = 1, 2).$$

Both polynomials F_1, F_2 have integral coefficients and F_2 is monic. If now $x_0, x_1, \ldots$ is a $P_1 - P_2$-cycle (finite or not), then one sees easily that $ax_0, ax_1, \ldots$ is an $F_1 - F_2$-cycle.

So assume that P_2 is monic and let first $x_0, x_1, \ldots, x_{r-1}$ be a finite $P_1 - P_2$-cycle. The preceding lemma shows that all x_i's are integers hence it suffices to establish the existence of a number $C = C(P_1, P_2)$ such that $|x_i| \leq C$ holds for $i = 0, 1, \ldots, r - 1$. To obtain this note first that $n < m$ implies that if $\max(|\alpha|, |\beta|)$ is sufficiently large and $P_2(\alpha) = P_1(\beta)$, then $|\beta| > |\alpha|$ must hold. In particular for sufficiently large $|x_i|$ we obtain $|x_{i-1}| > |x_i|$ and repeating this argument we are lead to

$$|x_0| = |x_r| < |x_{r-1}| < \cdots < |x_1| < |x_0|,$$

which gives a clear contradiction. Thus the union of all finite $P_1 - P_2$-cycles is bounded, hence finite.

Now let $\{x_i\}$ be an infinite $P_1 - P_2$-cycle. Using the preceding lemma we may assume that all its terms are integers, removing, if necessary, a finite number of its terms. As in the previous case we get $|x_{i-1}| > |x_i|$ provided $|x_i|$ is sufficiently large and hence for such i we get

$$0 \le |x_i| < |x_{i-1}| < \cdots < |x_1| < |x_0|,$$

which implies $|x_0| > i - 1$, a clear contradiction, since i may be arbitrarily large. $\square$

We return to the proof of the theorem. If $a \in A$, then one constructs a sequence $\{a_i\}$ of elements of A with $a_0 = a$ in the following way: if $a_0 = a, a_1, \ldots, a_s$ are already constructed, then choose for a_{s+1} any element of A satisfying $P_2(a_{s+1}) = P_1(a_s)$. If for some $0 \le i < j$ we have $a_i = a_j$ and i is smallest index with this property then we must have $i = 0$, since otherwise we would have

$$P_1(a_{i-1}) = P_2(a_i) = P_2(a_j) = P_1(a_{j-1}),$$

and the injectivity of P_1 on A gives $a_{i-1} = a_{j-1}$, contrary to the choice of i. This shows that the sequence $\{a_i\}$ is a finite or infinite $P_1 - P_2$-cycle. Thus A is contained in the union of all $P_1 - P_2$-cycles and the application of Lemma 11.4 concludes the proof. $\square$

3. The assertion of the preceding theorem holds also for other fields. K.K.KUBOTA [72a] (see also K.K.KUBOTA [72b],[73]) proved it for a class of fields encompassing all global fields, i.e. algebraic number fields and function fields in one variable over a finite field, and D.J.LEWIS [72] and P.LIARDET [70], [71], [72], [75] obtained its multidimensional analogue for all fields finitely generated over a prime field. We shall say that a field k has the *property* (**K**), if the following multidimensional analogue of Theorem 11.2 holds for k in place of **Q**:

Let $\Phi : k^n \longrightarrow k^n$ be an admissible polynomial mapping (as defined in Section 9) and let $\Psi : k^n \longrightarrow k^n$ be another polynomial mapping. Denote by d the minimum of the degrees of polynomials defining Φ and by D the maximum of the degrees of polynomials defining Ψ. If $d > D$, A is a subset of k^n satisfying $\Psi(A) \subset \Phi(A)$ and the restriction of Ψ to A is injective, then A is finite.

4. The following result of P.LIARDET [71] and D.J.LEWIS [72], which in one-dimensional case has been proved by K.K.KUBOTA [72a], contains the preceding theorem.

THEOREM 11.5. *All finite extensions of the rationals have the property* (**K**).

PROOF: Let $k/\mathbf{Q}$ be finite, let Φ and Ψ be as above and assume that $A \subset k^n$ satisfies $\Psi(A) \subset \Phi(A)$ and the restriction of Ψ to A is injective. For $a \in A$ define $T(a)$ as the unique element b of A satisfying $\Phi(b) = \Psi(a)$. From Lemma 10.6 one deduces that the Corollary to Lemma 9.2 is applicable with $f(a) = h(a)$, and the finiteness of A follows. $\square$

It has been shown in F.HALTER-KOCH,W.NARKIEWICZ [92a] that the property (K) is preserved under arbitrary finite extensions, as well as under purely transcendental extensions of arbitrary cardinality. This result implies in particular that all fields finitely generated over a prime field have the property (K). It is not known, whether (K) is equivalent to one of the properties considered in previous sections (**PROBLEM XX**).

Exercises

1. (K.K.KUBOTA [72a]) Give an example of polynomials $P_1, P_2 \in \mathbf{Q}[X]$ and an infinite subset A of $\mathbf{Q}$ satisfying

(i) deg $P_1 <$ deg P_2,
(ii) $P_1(A) = P_2(A)$,
   and
(iii) The restriction of P_2 to A is injective.

2. Let $k \geq 2$ be given. Show that if $P_1 \in \mathbf{Z}[X]$ and for all integers n the number $P_1(n)$ is a k-th power of an integer, then there exists a polynomial $P_2 \in \mathbf{Z}[X]$ such that
$$P_1(X) = P_2^k(X).$$

3. Let $P_1, P_2 \in \mathbf{Z}[X]$ and assume that $P_1(\mathbf{Z}) = P_2(\mathbf{Z})$ holds. Prove the existence of $a, b \in \mathbf{Z}$ such that
$$P_2(X) = P_1(aX + b).$$

4. (L.E.DICKSON [10]) Let p be an odd prime. Construct a polynomial over $\mathbf{F}_p$ whose all values in $\mathbf{F}_p$ are non-zero squares but which is not a square of a polynomial.

XII. Polynomial cycles

1. Recall that a finite subset $\{x_1, \ldots, x_n\}$ of a field K is called a *cycle* for f, if for $i = 1, 2, \ldots, n-1$ one has $f(x_i) = x_{i+1}$ and $f(x_n) = x_1$. The number n will be called the *length* of the cycle and the x_i's are called *cyclic elements of order n* or *fixpoints of f of order n*. We shall denote by $Cycl_K(f)$ the set of orders of all cyclic points of the polynomial f which lie in K.

Denote for $i = 1, 2, \ldots$ by f_i the i-th iterate of f, i.e. $f_1 = f$ and $f_{i+1} = f(f_i)$ for $i = 1, 2, \ldots$. Moreover put $f_0(X) = X$. The main properties of f_i are listed in the following lemma:

LEMMA 12.1. (i) *If* $\deg f = d$, *then* $\deg f_i = d^i$,

(ii) *For* $i, j = 1, 2, \ldots$ *one has* $f_{i+j} = f_i(f_j)$,

(iii) *If* $a \in K$, $f_n(a) = a$ *and* j *is the smallest integer satisfying* $f_j(a) = a$, *then* j *divides* n.

(iv) *Cyclic elements of order n of f coincide with those fixpoints of f_n which are not fixpoints of f_d where d runs over all proper divisors of n.*

(v) *If* $a \in K$, $f(a) = a$, *then for* $n = 1, 2, \ldots$ *one has*

(12.1)
$$f_n(X) = X + (f(X) - X)g_n(X),$$

with suitable polynomials $g_1, g_2, \ldots$ *satisfying*

$$g_n(a) = 1 + f'(a) + f'(a)^2 + \cdots + f'(a)^{n-1}.$$

PROOF: The properties (i) - (ii) follow immediately from the definitions.

To prove (iii) observe that if j does not divide n, then with suitable integers $N > 0$ and $0 < r < j$ we have $n = Nj + r$, thus using (ii) we get

$$a = f_n(a) = f_r(f_{Nj}(a)) = f_r \underbrace{(f_j(f_j(\ldots (f_j(a \ldots)))}_{N \text{ times}} = f_r(a),$$

contradicting the choice of j. Now (iv) follows immediately.

Finally to obtain (v) observe that

$$f(X) - f(Y) = (X - Y)F(X, Y),$$

where $F \in K[X,Y]$ and $F(X,X) = f'(X)$, thus for $n > 1$ we get

$$f_n(X) - f_{n-1}(X) = f(f_{n-1}(X)) - f(f_{n-2}(X))$$
$$= (f_{n-1}(X) - f_{n-2}(X))F(f_{n-1}, f_{n-2}),$$

and by recurrence we obtain with suitable $h_1, h_2, \ldots \in K[X]$,

$$f_n(X) - f_{n-1}(X) = (f(X) - X) \prod_{j=1}^{n-1} F(f_j(X), f_{j-1}(X)) = (f(X) - X)h_n(X).$$

Since

$$f_n(X) - X = f_n(X) - f_{n-1}(X) + \cdots + f_1(X) - X$$
$$= (f(X) - X)(1 + h_1(X) + \cdots + h_n(X)),$$

the equality (12.1) follows with $g_n = 1 + h_1 + \cdots + h_n$. To obtain the last assertion note that $f(a) = a$ implies $f_j(a) = a$ for all $j \geq 1$, thus

$$F(f_j(a), f_{j-1}(a)) = F(a, a) = f'(a),$$

and finally $h_j(a) = f'(a)^{n-1}$. $\square$

2. If K is algebraically closed then every nonlinear and non-constant polynomial over K has at least one cyclic point of order 1 in K. However in other fields there may exist polynomials without any cyclic points, as the example of the real field and the polynomial $x^2 + 1$ shows.

In an algebraically closed field every polynomial of degree ≥ 2 has many cyclic points:

THEOREM 12.2. (G.CHASSÉ [86]) *If K is an algebraically closed field and $f \in K[X]$ is of degree $d \geq 2$ then the set $Cycl_K(f)$ is infinite. More precisely, this set contains all prime numbers with at most d exceptions.*

PROOF: Let p be a prime number and assume that f does not have a cycle of order p. According to Lemma 12.1 (v) we have

$$f_p(X) - X = (f(X) - X)g_p(X),$$

and since $g_p \in K[X]$ is non-constant, it has a zero in K, say a, thus

$$0 = g_p(a) = 1 + f'(a) + \cdots + f'(a)^{p-1},$$

and we see that $f'(a)$ is a root of unity of order p.

Assume now that a is not a cyclic element of order p. Lemma 12.1 (iii) implies that in this case a must be a fixpoint of f. Thus we have at most d possibilites for a and it follows that p is the order of a root of unity of the form $f'(a)$, where a runs over a finite set of at most d elements. $\square$

3. More can be said in the case of complex numbers:

THEOREM 12.3. (I.N.BAKER [60]) *If f is a non-linear polynomial with complex coefficients then it has cyclic points of every order in the complex field, with at most one exception.*

PROOF: Let $k < n$ be positive integers and let f be a complex polynomial of degree $d > 1$ which does not have cyclic points of orders k and n. Clearly we must have $k > 1$.

Put

$$g(z) = \frac{f_n(z) - z}{f_{n-k}(z) - z}$$

and write $g(z) = P(z)/Q(z)$ with relatively prime polynomials P, Q. If q denotes the degree of Q, then Lemma 12.1 (i) implies $\deg P = d^n - d^{n-k} + q$, and it follows easily that the degree of the numerator of g' (in reduced form) does not exceed

$$c = d^n - d^{n-k} + 2q - 1,$$

hence g' has at most c zeros counted with their multiplicities.

We shall now obtain an upper bound for the numbers of solutions of $g(z) = 0$ and $g(z) = 1$:

LEMMA 12.4. (i) *The function g has at most*

$$\sum_{p|n} d^{n/p}$$

distinct zeros, with p running over all prime divisors of n,

(ii) *The equation $g(z) = 1$ has at most d^{n-1} distinct solutions.*

PROOF: (i) If a is a zero of g, then $f_n(a) = a$ and since by assumption f does not have cyclic points of order n, it follows from Lemma 12.1 (iii) that the minimal integer j satisfying $f_j(a) = a$ divides a maximal proper divisor of n. Since all maximal proper divisors of n are of the form n/p with prime p, we see that a is a zero of $f_{n/p}$ for a suitable prime p, and this implies the assertion.

(ii) If $g(a) = 1$ then by Lemma 12.1 (ii) we get

$$f_{n-k}(a) = f_n(a) = f_k(f_{n-k}(a)),$$

hence $f_{n-k}(a)$ is a fixpoint of f_k. Since f does not have cyclic points of order k, $f_{n-k}(a)$ must be a fixpoint of f_j for a certain proper divisor j of k, i.e.

$$f_j(f_{n-k}(a)) = f_{n-k}(a),$$

and Lemma 12.1 (ii) shows now that a is a zero of $f_{n-k+j} - f_{n-k}$. The degree of the last polynomial being equal to d^{n-k+j} we obtain for the number of possible a's the bound

$$\sum_{\substack{j|k \\ j<k}} d^{n-k+j},$$

which in case $k = 2$ equals d^{n-1} and in case $k > 2$ does not exceed

$$\sum_{j=1}^{k-2} d^{n-k+j} = d^{n-k+1} \frac{d^{k-2} - 1}{d - 1} < d^{n-1}. \quad \square$$

COROLLARY. *The total number of solutions of $g(x) = 0$ and $g(x) = 1$ counted with their multiplicities does not exceed*

$$d^n - d^{n-k} + d^{n-1} + 2q - 1 + \sum_{p|n} d^{n/p}.$$

PROOF: Use both parts of Lemma 12.4 and the fact that if a is a zero of $g(x)$ or $g(x) - 1$ of order N, then it is also a zero of $g'(x)$ of order $N - 1$, and as we have seen above, g' has at most $c = d^n - d^{n-k} + 2q - 1$ zeros, the multiplicities counted. $\square$

Now we can conclude the proof of Theorem 12.3. The total number of solutions of $g(x) = 0$ and $g(x) = 1$, counted with their multiplicities, equals $2(d^n - d^{n-k} + q)$, thus the Corollary implies

$$2(d^n - d^{n-k} + q) \le d^n - d^{n-k} + d^{n-1} + 2q - 1 + \sum_{p|n} d^{n/p},$$

and we arrive at

$$1 + d^n \le \sum_{p|n} d^{n/p} + d^{n-k} + d^{n-1}.$$

The sum $\sum_{p|n} d^{n/p}$ equals d^{n-2} for $n = 3, 4$ and is bounded by

$$1 + d + \cdots + d^{n-3} < d^{n-2}$$

for $n \ge 5$. Thus finally we obtain

$$d^n < 1 + d^n \le 2d^{n-2} + d^{n-1},$$

which is not possible for $d \ge 2$. $\square$

4. The exceptions can really occur, as the example $P(X) = X^2 - X$ shows. One sees easily that in this case there are no cyclic points of order 2.

We show now that the assertion of the theorem holds for all algebraically closed fields of zero characteristics:

COROLLARY. *If P is a non-linear polynomial over an algebraically closed field K of zero characteristics, then the set $Cycl_K(P)$ contains all positive integers with at most one exception.*

PROOF: If k is the field generated over the rationals by the coefficients of P then it is finitely generated and hence has an isomorphic copy contained in $\mathbf{C}$. Thus its algebraic closure L can be also regarded as a subfield of $\mathbf{C}$ and it follows that P can be regarded as a polynomial with complex coefficients. The theorem shows now that P has complex cyclic points of every order, with at most one exception. These points, being obviously algebraic over k, must lie in $L \subset K$ and this establishes our assertion. $\square$

We show now that this corollary does not hold in non-zero characteristics:

THEOREM 12.5. *If p is a prime and K is an algebraically closed field of characteristics p, then the polynomial*

$$f(X) = X^p + X$$

has no cycles of lengths p^k for $k = 1, 2, \ldots$ in K.

PROOF: It suffices to consider the case when K is the algebraic closure $\hat{\mathbf{F}}_p$ of $\mathbf{F}_p$, since all elements of an f-cycle are algebraic over $\mathbf{F}_p$.

LEMMA 12.6. *If f_j denotes the j-th iterate of f then for $j = 1, 2, \ldots$ one has*

$$f_j(X) = x^{p^j} + \sum_{r=1}^{j-1} \binom{j}{r} x^{p^r} + x.$$

PROOF: The assertion being evident in case $j = 1$ assume its truth for a certain j and observe that in view of $\binom{j}{r}^p = \binom{j}{r}$ and $(a+b)^p = a^p + b^p$ we get

$$f_{j+1}(X) = f(f_j(X)) = (x^{p^j} + \sum_{r=1}^{j-1} \binom{j}{r} x^{p^r} + x)^p + x^{p^j} + \sum_{r=1}^{j-1} \binom{j}{r} x^{p^r} + x$$

$$= x^{p^{j+1}} + \sum_{r=1}^{j-1} \binom{j}{r} x^{p^{r+1}} + x^p + x^{p^j} + \sum_{r=1}^{j-1} \binom{j}{r} x^{p^r} + x$$

$$= x^{p^{j+1}} + (\binom{j}{j-1} + 1) x^{p^j} + \sum_{r=2}^{j-1} (\binom{j}{r-1})$$

$$+ \binom{j}{r}) x^{p^r} + (\binom{j}{1} + 1) x^p + x$$

$$= x^{p^{j+1}} + \sum_{r=1}^{j} \binom{j+1}{r} x^{p^r} + x,$$

as asserted. $\square$

COROLLARY. *For $s = 1, 2, \ldots$ one has*

$$f_{p^s}(X) = X^{p^{p^s}} + X.$$

PROOF: It suffices to note that $\binom{p^s}{r}$ is divisible by p for $r = 1, 2, \ldots, p^s - 1$. $\square$

The theorem follows now easily: if $\xi \in \hat{\mathbf{F}}_p$ lies in a cycle of length p^s then $f_{p^s}(\xi) = \xi$ and the corollary implies $\xi = 0$, i.e. ξ is a fixpoint of f. $\square$

A study of cycles for a large class of polynomials over the algebraic closure of a finite field is contained in a paper of A.BATRA and P.MORTON [94]. Recently T.PEZDA [94b] considered the orders of cyclic points of polynomials in algebraically closed fields of characteristics $p > 0$ and determined all polynomials which have cyclic points of all orders with at most finitely many exceptions. He proved in particular that for polynomials with degree not divisible by p there are at most 8 exceptional cycle-lengths.

Note that the analogue of Theorem 12.2 fails in higher dimensions, as e.g. the map $T : \mathbf{C}^2 \longrightarrow \mathbf{C}^2$ defined by $T : (x, y) \mapsto (y^2 + x + 2, y^2)$ has no cyclic points at all. Indeed, if (a, b) lies in a cycle of length n then one gets easily

$$2n + b + b^2 + b^4 + \cdots + b^{2^{n-2}} + b^{2^{n-1}} = 0$$
$$b^{2^{n-1}} = 1,$$

which leads to $|b| = 1$ making the first equality impossible.

5. A study of cycles of rational functions in the complex plane has been started by D.FATOU [19] who later (D.FATOU [26]) considered this question for entire transcendental functions and obtained the analogue of Theorem 12.2 in this case. In a special case this has been made more precise by I.N.BAKER [59a], who later showed (I.N.BAKER [60]) that in Theorem 12.3 one may replace polynomials by arbitrary non-linear entire functions. Later, using stronger methods, I.N.BAKER [64] has been able to prove that the exceptional cycle-length appearing in Theorem 12.3 equals 2 and occurs only for quadratic polynomials, linearly conjugated to $X^2 - X$. (Two polynomials f, g are called *linearly conjugated* if there exists a linear polynomial L such that $L(f) = g(L)$). This assertion fails for transcendental entire functions, since the function $e^z + z$ does not have fixpoints of order 1, however I.N.BAKER [60] conjectured that it holds true with 2 replaced by 1 and showed (I.N.BAKER [59b]) that in certain cases no exceptions can occur. This conjecture has been recently established by W.BERGWEILER [91]. The same assertion holds also for meromorphic transcendental functions (P.BHATTACHARYYA [69], W.BERGWEILER [93]). The last paper presents a survey of results of this type dealing with meromorphic transcendental functions.

The case of rational functions has been studied in I.N.BAKER [64] who has shown that if d is the maximum of degrees of numerator and denominator in the representation of a rational function $f \in \mathbf{C}(X)$ as a ratio of two polynomials then f has cyclic points of every order N with the possible exception of the following three cases:

(a) $d = 1$,
(b) $d = 2$; $N = 2, 3$.
(c) $d = 3, 4$; $N = 2$.

The structure of orbits of rational maps over an algebraic number field has been considered by J.SILVERMAN [93]. He proved a.o. that if f is such a map which is not a homography and $f(f(X))$ is not a polynomial, then every orbit can contain only finitely many integral points.

For real polynomials the situation is much more complicated. It is immediate that the polynomial X^2 does not have real fixpoints of order $\neq 1$ and the polynomial $X^2 + X + 1$ does not have real fixpoints of any order. In fact it is easy to see that if g is a real polynomial without real zeros, then the polynomial $f(X) = g(X) + X$ has no real fixpoints of any order.

6. In the remaining part of this section we show that the length of a cycle of a monic polynomial f with integral algebraic coefficients is bounded by a value which depends only on the degree of the field generated by coefficients of f. We start with a few general observations. Thus let R be an arbitrary domain and denote by K its field of fractions.

The following lemma reduces the study of possible cycle-lengths to certain particular cases:

LEMMA 12.7. *Let R be a domain and assume that the polynomial $f \in R[X]$ has a cycle of length n in R.*

(i) *If d divides n, then the n/d-th iterate of f has in R a cycle of length d.*

(ii) *There exists a monic polynomial f in $R[X]$ having a cycle of length n which starts with the elements 0 and 1.*

PROOF: The assertion (i) is evident and to prove (ii) let $x_0, x_1, \ldots, x_{n-1} \in R$ be a cycle of f. Then the polynomial $F(X) = f(X + x_0) - x_0 \in R[X]$ has the cycle $0, y_1, \ldots, y_{n-1}$ with $y_i = x_i - x_0$ $(i = 1, 2, \ldots, n-1)$.

Observe now that every y_i is divisible by y_1. In fact, for $i = 1$ this is clear, and if $y_1, y_2, \ldots, y_{i-1}$ are divisible by y_1, and

$$F(X) = a_N X^N + a_{N-1} X^{N-1} + \cdots + a_1 X + a_0$$

then $a_0 = y_1$ and

$$y_i = F(y_{i-1}) = a_N y_{i-1}^N + a_{N-1} y_{i-1}^{N-1} + \cdots + a_1 y_{i-1} + y_1 \equiv 0 \pmod{y_1}.$$

Thus the polynomial $g(X) = F(X y_1)/y_1 \in R[X]$ has the cycle $0, 1, z_2, \ldots, z_{n-1}$ with $z_i = y_i/y_1 \in R$ $(i = 2, 3, \ldots, n-1)$ and finally the polynomial

$$g(X) + X^k(X - 1) \prod_{j=2}^{n-1} (X - z_j) \in R[X]$$

is monic for sufficiently large k and has the cycle $0, 1, z_2, \ldots, z_{n-1}$. $\square$

On the basis of this lemma we may assume henceforth that the cycles considered start with 0,1.

The next lemma collects certain properties of cycles on which further results will be based:

LEMMA 12.8. *Let R be a domain and $f \in R[X]$ a monic polynomial of degree $d > 1$. Let $x_0 = 0, x_1 = 1, x_2, \ldots, x_{n-1}$ be a cycle for f of length $n \geq 2$ lying in R and define x_k for $k \geq n$ by putting $x_k = x_{k \bmod n}$. Then for any fixed $k > 0$ not divisible by n the differences $x_{j+k} - x_j$ are associated in R (i.e. their ratio is* *invertible) for $j = 0, 1, \ldots$. In particular all differences $x_{j+1} - x_j$ are invertible in R.*

PROOF: Note that for all i we have

$$x_{i+k} - x_i = P(x_{i+k-1}) - P(x_{i-1}),$$

hence

$$x_{i+k-1} - x_{i-1} \mid x_{i+k} - x_i,$$

and we get

$$x_k = x_k - x_0 \mid x_{k+1} - x_1 \mid \ldots \mid x_{k+n} - x_n = x_k,$$

which implies the first part of the lemma. The second part results now from the equality $x_1 - x_0 = 1$. $\square$

COROLLARY 1. *Under the assumptions and notation of the lemma we have:*

(i) *If k divides m, then x_k divides x_m,*

(ii) *If $(k, n) = 1$, then x_k is invertible in R. In particular x_{n-1} is invertible.*

(iii) *If n is a prime, then all elements x_i and $x_i - x_j$ are for $i, j = 1, 2, \ldots, n$ $(i \neq j)$ invertible in R.*

PROOF: (i) The lemma implies the divisibility of all elements $x_{2k} - x_k, x_{3k} - x_{2k}, \ldots$ by x_k and thus $x_{2k}, x_{3k}, \ldots$ are also divisible by x_k.

(ii) If $km \equiv 1 \pmod{n}$ then $x_{km} = x_1 = 1$ and by (i) x_k divides 1.

(iii) The first assertion follows from (ii) and to obtain the second observe that $x_i - x_j$ is associated with x_{i-j}. $\square$

For any ring R one defines its *Lenstra constant* $L(R)$ as the largest cardinality of a subset A of R with the property that for all distinct $a, a' \in A$ the difference $a - a'$ is invertible. (This constant has been used by H.W.LENSTRA [77] in his study of Euclidean rings of algebraic numbers. For the case of arbitrary rings see A.LEUTBECHER,G.NIKLASCH [89]).

COROLLARY 2. *If R is a domain with finite Lenstra constant $L(R)$ and n is the length of a polynomial cycle in R, then no prime divisor of n can exceed $L(R)$.*

PROOF: Apply part (iii) of the preceding Corollary. $\square$

It follows easily from the lemma that if $y_0, y_1, \ldots, y_{n-1} \in R$ form a cycle of length n, then the ratios $(y_{j+k} - y_j)/(y_k - y_0)$ are all invertible. They have been called *dynamical units* and studied in P.MORTON, J.H.SILVERMAN [93a].

7. We shall now consider finite extensions $K/\mathbf{Q}$ of the rationals and give a bound for the length of polynomial cycles in the ring $\mathbf{Z}_K$ of integers of K. We look first at the simplest case $K = \mathbf{Q}$:

THEOREM 12.9. (W.NARKIEWICZ [89]) *A monic polynomial with rational integral coefficients can have in* $\mathbf{Q}$ *only cycles of length 1 or 2.*

PROOF: As $\mathbf{Z}$ is integrally closed, all finite cycles of a monic $\mathbf{Z}$-polynomial lie necessarily in $\mathbf{Z}$. If there exists such a polynomial f with a cycle $x_0 = 0, x_1 = 1, \ldots, x_{n-1}$ with $n > 2$ then by Lemma 12.8 we have $x_{i+1} - x_i = \pm 1$ and in view of $n > 2$ Corollary 1 (ii) to that lemma gives $x_{n-1} = -1$. Hence one can pass from $x_1 = 1$ to $x_{n-1} = -1$ using steps equal to ± 1 without passing through $x_0 = 0$ which gives a contradiction. $\square$

The same argument is applicable in every domain in which ± 1 are the only invertible elements. In particular this holds for rings of integers in imaginary quadratic fields, distinct from $\mathbf{Q}(i)$ and $\mathbf{Q}(\sqrt{-3})$.

It has been observed by G.BARON (in a letter to the author) that in case $R = \mathbf{Z}$ the number of cycles of a given length ≤ 2 cannot be arbitrarily prescribed. He proved namely the following result:

THEOREM 12.10. *A monic polynomial with rational integral coefficients has in* $\mathbf{Z}$ *either no cycles at all or one cycle of order 1 and $k \geq 0$ cycles of order 2 or finally $k \geq 1$ cycles of the same order (either 1 or 2).*

PROOF: Assume that a monic polynomial $f(X) \in \mathbf{Z}[X]$ has in $\mathbf{Z}$ a cycle of length 2, which due to Lemma 12.7 (ii) may be assumed to be $0, c$ and two distinct fixpoints, say $a \neq b$. Obviously $abc \neq 0$. We can write

$$f(X) = (X - a)(X - b)R(X) + X$$

with a suitable $R \in \mathbf{Z}[X]$ and in view of $c = f(0) = abR(0)$ we get $R(0) \neq 0$. Moreover

$$0 = f(c) = (c - a)(c - b)R(c) + c = ab\left((bR(0) - 1)(aR(0) - 1)R(c) + R(0)\right)$$

leading to $0 = (bR(0) - 1)(aR(0) - 1)R(c) + R(0)$ which shows that the numbers $R(c)$ and $R(0)$ divide each other, thus $R(c) = \pm R(0) \neq 0$. This gives $0 = (bR(0) - 1)(aR(0) - 1) \pm 1$ and finally we get $bR(0) - 1 = \pm 1$ and $aR(0) - 1 = \pm 1$, thus, in view of non-vanishing of $aR(0)$ and $bR(0)$ gives $aR(0) = bR(0) = 2$, implying $a = b$, a contradiction. $\square$

8. In the case of an arbitrary algebraic number field and a not necessarily monic polynomial things become more complicated:

THEOREM 12.11. (W.NARKIEWICZ [89]) *Let K be an algebraic number field of degree M, $\mathbf{Z}_K$ its ring of integers and let $f \in K[X]$ be a non-constant polynomial. Write*

$$f(X) = \frac{1}{D} \sum_{j=0}^{N} A_j X^j,$$

with $A_0, A_1, \ldots, A_N, D \in \mathbf{Z}_K$ ($A_N \neq 0$), let $\mathcal{P}$ be the set of all prime ideals of $\mathbf{Z}_K$ which divide $DA_N \mathbf{Z}_K$ and denote by T the cardinality of $\mathcal{P}$. If $r_1, 2r_2$

are the numbers of real resp. non-real embeddings of K in the field of complex numbers then put

$$w = w(f) = r_1 + r_2 + T.$$

There exists a number $B = B(w, M)$ such that the length of every cycle of f lying in K is bounded by B.

PROOF: To prove our result we have to appeal to the following deep result of J.H.EVERTSE [84] concerning diophantine equations and which we shall quote here without proof. To state it we recall first two notions from algebraic number theory:

Let S be a finite set of inequivalent absolute values on K, containing all archimedean absolute values. An element a of K is called S-integral, provided for all absolute values $v \notin S$ we have $v(a) = 1$ and it is called an S-unit if both a and $1/a$ are S-integral.

LEMMA 12.12. (Theorem of Evertse) For any non-zero element $a \in K$ the number of solutions of the equation

$$x + y = a,$$

(with x and y being S-units) does not exceed $3 \cdot 7^{M+2s}$, where s denotes the cardinality of S.

To apply this result we take for S the set of all these absolute values of K which are either archimedean or correspond to prime ideals in $\mathcal{P}$. Let R be the ring of all S-integral elements of K, thus invertible elements of R coincide with S-units. Observe that all coefficients of f lie in R and its leading coefficient is an S-unit. Since R is integrally closed, all finite cycles of f in K lie in R. If n is the length of such a cycle and p is a prime divisor of n, then Corollary 2 to Lemma 12.8 shows that p cannot exceed the Lenstra constant $L(R)$ of R, and thus

$$p \leq L(R) \leq t(\mathcal{P}),$$

where $t(\mathcal{P})$ denotes the smallest norm of a prime ideal not in $\mathcal{P}$.

The next lemma establishes a bound for $t(\mathcal{P})$ depending only on the degree M of K and T.

LEMMA 12.13. One has

$$t(\mathcal{P}) \leq (cT)^M \log^M T,$$

where c is an absolute constant.

PROOF: Let $p_1 = 2 < p_2 < \ldots$ be the sequence of all rational primes. Since every prime ideal $P \in \mathcal{P}$ contains exactly one rational prime, say p, and $N(P) = p^f$ with a suitable $f \leq M$, the inequality

$$t(\mathcal{P}) \leq p_{1+T}^M$$

follows and it suffices to recall the inequality $p_k \leq c_1 k \log k$ valid, with a suitable c_1, for $k = 1, 2, \ldots$. $\square$

COROLLARY. *If p is a prime dividing the length of a f-cycle in R, then p cannot exceed $(cT)^M \log^M T$.* $\square$

Now we bound the prime power divisors of n.

LEMMA 12.14. *If p^m is a prime power dividing the length of an f-cycle in R and $m > 1$ then*
$$p^m \leq 6 \cdot 7^{M+2w(f)}.$$

PROOF: By Lemma 12.7 (i) it suffices to consider only the case of a cycle $x_0 = 0, x_1 = 1, x_2, \ldots$ whose length is a prime power p^m. Put $s = p^2 - p$ and observe that for k satisfying $0 < k < p^m$ and not divisible by p both terms on the right hand-side of the equality
$$x_s = x_{s-k} + (x_s - x_{s-k})$$
are invertible in R in view of Lemma 12.8 and part (ii) of the corollary to that lemma. Hence the equation
$$x_s = u_1 + u_2$$
has at least $\varphi(p^m) = p^{m-1}(p-1)$ solutions with invertible u_1, u_2 and the application of Lemma 12.12 leads to
$$p^m \leq 2p^{m-1}(p-1) \leq 6 \cdot 7^{M+2w(f)},$$
as asserted. $\square$

The theorem follows now readily: Corollary to Lemma 12.13 shows that all prime divisors of a cycle-length n are bounded by a number, depending only on $w(f)$ and M and Lemma 12.14 establishes the same for all prime powers $p^m > p$ dividing n. Hence n itself is bounded by a similar number. $\square$

9. The possible cycle-lengths in rings of integers in quadratic number fields have been determined independently by J.BODUCH [90] in his M.A. thesis (in which however the question of the existence of a cycle of length 6 in the ring of integers of $\mathbf{Q}(\sqrt{5})$ has been left open) and by G.BARON (in a letter to the author). The result is as follows:

Let $K = \mathbf{Q}(\sqrt{d})$ with a square-free d and let R_d be the ring of integers of K. If $d \neq -3, -1, 2, 5$ then the only possible cycle-lengths in R_d equal 1 and 2. For $d = -1, 2$ there are cycles of length $1, 2$ and 4, for $d = -3$ there are cycles of length $1, 2, 3$ and 6 and for $d = 5$ there are cycles of length $1, 2, 3$ and 4. These lists are exhaustive.

For fields K of larger degree the problem of determining all cycle-lengths in their rings of integers $\mathbf{Z}_K$ is still open (**PROBLEM XXI**).

Cycles of quadratic polynomials have been recently studied by P.MORTON [92] and P.RUSSO, R.WALDE [94]. In P.MORTON [92,II] it has been shown that quadratic polynomials with rational coefficients cannot have cycles of length four in $\mathbf{Q}$.

10. Recently T.PEZDA [93] gave a fresh proof of Theorem 12.11 which does not use the result of Evertse and is based instead on p-adic considerations. He showed that the lengths of cycles of polynomial mappings in a discrete valuation domain R of zero characteristics with a finite residue field are bounded by a number depending only on R, and this applies in particular to rings of integral p-adic numbers. This leads to an essentially better bound for cycle-lengths in rings of algebraic integers than that resulting from the proof of Theorem 12.11, namely $2^{M+1}(2^M - 1)$ with M being the degree of the field. (Similar results have been independently obtained by P.MORTON, J.H. SILVERMAN [93b], who considered maps defined by rational functions).

In the case of $\mathbf{Z}_p$, the ring of p-adic integers, Pezda showed that an integer is the length of a polynomial cycle if and only if it can be written as ab with $1 \leq a \leq p$ and $b \mid p - 1$, except in the cases $p = 2, 3$ in which additionally the cycles of length p^2 occur. Later he extended (T.PEZDA [94a]) his result to polynomial maps in several variables and in particular showed that any polynomial map in N variables over the ring of integers $\mathbf{Z}_K$ in an algebraic number field K has the lengths of its cycles in $\mathbf{Z}_K^N$ bounded by $2^{n(1+3N+N^2)}$ where $n = [K : \mathbf{Q}]$. He obtained also a corresponding result for discrete valuation rings of zero characteristics and finite residue field. A similar result for all finitely generated domains of zero characteristics has been obtained by F.HALTER-KOCH, W.NARKIEWICZ [94].

Exercises

1. (i) Show that the rational functions $X - 2(X^3 - 1)/3X^2$ and $-X(1 + 2X^3)/(1 - 3X^3)$ do not have cycles of order 2 in the complex plane.

(ii) Show that the function $X + (\zeta_3 - 1)(X^2 - 1)/2X$ (where ζ_3 denotes the third primitive root of unity) does not have cycles of order 3 in the complex plane.

2. Let R be a domain with quotient field K and $f \in R[X]$. Show that if f has a cycle $x_0 = 0, x_1 = 1, x_2, \ldots, x_{n-1}$ then the Lagrange interpolation polynomial $L \in K[X]$ for the data $L(x_i) = x_{i+1}$ $(i = 0, 1, \ldots, n - 2)$, $L(x_{n-1}) = 0$ has its coefficients in R.

3. Let R be a domain. Prove that there is a polynomial in R with a cycle of length 3 if and only if the equation

$$u + v = 1$$

has a solution with u, v being invertible elements of R.

4. (G.BARON) (i) Assume that $f \in \mathbf{Z}[X]$ has two 2-element cycles $\{a_0, a_1\}$ and $\{b_0, b_1\}$ in $\mathbf{Z}$. Show that $a_0 + a_1 = b_0 + b_1$.

(ii) Assume that $f \in \mathbf{Z}[X]$ has a 2-element cycle $\{a_0, a_1\}$ in $\mathbf{Z}$ and a fixpoint $b \in \mathbf{Z}$. Prove $a_0 + a_1 = 2b$.

(iii) Show that for every $k \geq 0$ there exists a monic polynomial in $\mathbf{Z}[X]$ with one fixpoint and k cycles of length 2 in $\mathbf{Z}$.

(iv) Show that for every $k \geq 0$ there exists a monic polynomial in $\mathbf{Z}[X]$ with k fixpoints and no other cycles in $\mathbf{Z}$.

5. Let K be a quadratic number field, $\mathbf{Z}_K$ its ring of integers and let $f \in \mathbf{Z}_K[X]$.

(i) Prove that if $K = \mathbf{Q}(i)$ or $K = \mathbf{Q}(\sqrt{2})$, then f has only cycles of lengths 1,2 or 4.

(ii) Prove that if $K = \mathbf{Q}(\sqrt{-3})$ then f can have only cycles of lengths 1,2,3 or 6.

(iii) Prove that if $K = \mathbf{Q}(\sqrt{3})$ then f can have only cycles of lengths 1 or 2.

6. Let p be a rational prime. Show that for every $n = 1, 2, \ldots, p-1$ there is a polynomial f with coefficients in the ring $\mathbf{Z}_p$ of p-adic integers which has in $\mathbf{Z}_p$ a cycle of length n.

7. (i) Let $f \in \mathbf{Z}[X]$ be monic, let p be a prime, $c \geq 0$ and assume that $x_0 = 0, x_1 = 1, x_2, \ldots, x_{n-1}$ is a cycle for f in the ring of residue classes mod p^{1+2c}. Assume further that the product

$$\prod_{i=0}^{n-1} f'(x_i)$$

is congruent to unity mod p^c but not mod p^{1+c}. Prove that f has a cycle of n elements in $\mathbf{Z}_p$.

(ii) Prove that for every p there is a polynomial $f \in \mathbf{Z}_p[X]$ having a cycle of length p in $\mathbf{Z}_p$.

(iii) Give an example of a polynomial having cycles of length 4 in $\mathbf{Z}_2$.

(iv) Prove that if n is a cycle-length in $\mathbf{Z}_p$, then n cannot have a prime divisor exceeding p.

(v) Prove that if there is a polynomial over $\mathbf{Z}_p$ with a cycle of n elements, then there exists a polynomial over $\mathbf{Z}_p \cap \mathbf{Q}$ which has a cycle of the same length in $\mathbf{Z}_p \cap \mathbf{Q}$.

8. (P.MOUSSA, J.S.GERONIMO, D.BESSIS [84]). Let $f(X) \in \mathbf{Z}[X]$ be monic and let a be an algebraic integer. Prove that if for every conjugate a' of a the sequence $\{a', f(a'), f_2(a'), \ldots\}$ is bounded, then for a suitable n the number $f_n(a)$ lies in a finite cycle for f.

List of open problems

We give here a list of open problems which appear in the main text, where attributions and comments may be found.

REFERENCES

BOOKS:

[A] van der WAERDEN,B.L. *Algebra*, 5th ed., Springer 1967. MR 38#1968.

[B] BOURBAKI,N. *Algèbre commutative, VII, Diviseurs*, Hermann, 1965. MR 41#5339.

[CF] CASSELS,J.W.S., FRÖHLICH,A. (editors), *Algebraic Number Theory*, Academic Press, 1967. MR 35#6500.

[DG] LANG,S. *Fundamentals of Diophantine Geometry*, Springer 1983. MR 85j:11005.

[EATAN] NARKIEWICZ,W. *Elementary and Analytic Theory of Algebraic Numbers*, Springer & PWN, 1990. MR 91h:11107.

[H] HASSE,H. *Zahlentheorie*, Akademie Verlag 1949. MR 11 p.580.

[IRF] BEARDON,A.F. *Iteration of rational functions*, Springer 1991. MR 92j:30026.

[KR] KRONECKER L. *Vorlesungen über Zahlentheorie* I, Leipzig 1901.

[MD] McDONALD,B.R. *Finite rings with identity*, M.Dekker 1974. MR 50#7245.

[MIT] GILMER,R. *Multiplicative Ideal Theory*. M.Dekker, 1972. MR 55#323.

[SH] SHAFAREVITCH,I.R. *Fundamentals of Algebraic Geometry*, 2nd ed., Moscow 1988. MR 90g:14001–14002. (Russian)

PAPERS:

ACZEL,J. [60] *Ueber die Gleichheit der Polynomfunctionen auf Ringen*, Acta Sci. Math., 21, 1960, 105–107. MR 24#A1291a.

ANDERSON,D.D.,ANDERSON,D.F.,ZAFRULLAH,M. [91] *Rings between D[X] and K[X]*, Houston J.Math., 17, 1991, 109–129. MR 92c:13014.

ANDERSON,D.F., BOUVIER,A., DOBBS,D.E., FONTANA,M., KABBAJ, S. [88] *On Jaffard domains*, Expos. Math., 5, 1988, 145–175. MR 89c:13014.

ARTIN,E.,WHAPLES,G. [45] *Axiomatic characterization of fields by the product formula for valuations*, Bull. Amer. Math. Soc., 51, 1945, 469–492. MR 7 p.111.

AVANISSIAN,V.,GAY,R. [75] *Sur une transformation des fonctionnelles analytiques et ses applications aux fonctions entières de plusieurs variables*, Bull. Soc. Math. France, 103, 1975, 341–384. MR 53#848.

BAKER,A. [67] *A note on integral integer-valued functions of several variables*, Proc. Cambridge Phil. Soc., 63, 1967, 715–720. MR 35#4168.

BAKER,I.N. [59a] *Fixpoints and iterates of entire functions*, Math. Zeitschr., 71, 1959, 146–153. MR 21#5743.

— [59b] *Some entire functions with fixpoints of every order*, J. Austral. Math. Soc. 1, 1959/61, 203–209. MR 22#4838a.

— [60] *The existence of fixpoints of entire functions*, Math. Zeitschr., 73, 1960, 280–284. MR 22#4838b.

— [64] *Fixpoints of polynomials and rational functions*, J. London Math. Soc., 39, 1964, 615–622. MR 30#230.

BARSKY,D. [72] *Polynômes dont les dérivées sont à valeurs entières et fonctions k-lipschitziennes sur un anneau local*, Thése 3 cycle, Univ. Paris VII, 1972.

— [73] *Fonctions k-lipschitziennes sur un anneau local et polynômes à valeurs entiéres*, Bull. Soc. Math. France, 101, 1973, 397–401. MR 51#8080.

BASS,H. [63] *Big projective modules are free*, Illinois J. Math., 7, 1963, 24–31. MR 26#1341.

BATRA,A.,MORTON,P. [94] *Algebraic dynamics of polynomial maps on the algebraic closure of a finite field,I*, Rocky Mount. J. Math., 24, 1994, 453–481.

BENZAGHOU,B. [70] *Algébres de Hadamard*, Bull. Soc. Math. France, 98, 1970, 209–252. MR 44#1658.

BERGWEILER,W. [91] *Periodic points of entire functions: proof of a conjecture of Baker*, Complex Variable Theory Appl., 17, 1991, 57–72.

—. [93] *Iteration of meromorphic functions*, Bull. Amer. Math. Soc., 29, 1993, 151–188.

BÉZIVIN,J.-P. [84] *Une généralisation à plusieurs variables d'un résultat de Gel'fond*, Analysis, 4, 1984, 125–141. MR 86f:11053.

BODUCH,J. [90] *Polynomial cycles in rings of algebraic integers*, MA thesis, Wroclaw University 1990. (Polish)

BRAWLEY,J.V., MULLEN,G.L. [92] *Functions and polynomials over Galois rings*, J. Number Th., 41, 1992, 156–166. MR 93d:11124.

BREWER,J.,KLINGLER L. [91] *The ring of integer-valued polynomials of a semi-local principal-ideal domain*, Lin. Alg. Appl., 15, 1991, 141–145. MR 92g:13025.

BRIZOLIS,D. [74] *On the ratios of integer-valued polynomials over any algebraic number field*, American Math. Monthly, 81, 1974, 997–999. MR 50#7085.

— [75] *Hilbert rings of integer-valued polynomials*, Comm. Algebra, 3, 1975, 1051–1081. MR 52#10697.

— [76] *Ideals in rings of integer-valued polynomials*, J. reine angew. Math., 285, 1976, 28–52. MR 54#2643.

— [79] *A theorem on ideals in Prüfer rings of integer-valued polynomials*, Comm. Algebra, 7, 1979, 1065–1077. MR 80j:13013.

BRIZOLIS,D., STRAUS,E.G. [76] *A basis for the ring of double integer-valued polynomials*, J. reine angew. Math., 286/287, 1976, 187–195. MR 54#5168.

de BRUIJN,N.G. [55] *Some classes of integer-valued functions*, Indag. Math., 17, 1955, 363–367. MR 17 p.128.

BUCK,C.R. [48] *Integral valued entire functions*, Duke Math. J., 15, 1948, 879–891. MR 10 p.693.

BUNDSCHUH,P. [80] *Arithmetische Eigenschaften ganzer Funktionen mehreren Variablen*, J. reine angew. Math., 313, 1980, 116–132. MR 81i:10037.

CAHEN,P.-J. [72] *Polynômes à valeurs entières*, Canad. J. Math., 24, 1972, 747–754. MR 46#9027.

— [75] *Polynômes et dérivées à valeurs entiers*, Ann. Sci. Univ. Clermont II, sér. math., 10, 1975, 25–43. MR 53#388.

— [78] *Fractions rationnelles à valeurs entières*, Ann. Sci. Univ. Clermont II, sér. math., 16, 1978, 85–100. MR 80e:12001.

— [90] *Dimension de l'anneau des polynômes à valeurs entières*, Manuscr. Math., 67, 1990, 333–343. MR 91e:13010.

— [91] *Polynômes à valeurs entières sur un anneau non analytiquement irréductible*, J. reine angew. Math., 418, 1991, 131–137. MR 92i:13015.

— [93] *Parties pleines d'un anneau noethérien*, J. Algebra, 157, 1993, 199–212.

CAHEN,P.-J., CHABERT, J.-L. [71] *Coefficients et valeurs d'un polynôme*, Bull. Sci. math., (2), 95, 1971, 295–304. MR 45#5126.

CAHEN,P.J.,HAOUAT,Y. [88] *Polynômes à valeurs entières sur un anneau de pseudovaluation*, Manuscr. Math., 61, 1988, 23–31. MR 89f:13007.

CARLITZ,L. [59] *A note on integral-valued polynomials*, Indag. Math., 21, 1959, 294–299, MR 21#7178.

— [63] *Permutations in a finite field*, Acta Sci. Math., 24, 1963, 196–203. MR 28#81.

— [64] *Functions and polynomials* (mod p^n), Acta Arith., 9, 1964, 67–78. MR 29#4721.

CARLSON,F. [21] *Über ganzwertige Funktionen*, Math. Zeitschr., 11, 1921, 1–23.

CHABERT,J.-L. [71] *Anneaux de "polynômes à valeurs entières" et anneaux de Fatou*, Bull. Soc. Math. France, 99, 1971, 273–283. MR 46#1780.

— [77] *Les idéaux premiers de l'anneau des polynômes à valeurs entières*, J. reine angew. Math., 293/294, 1977, 275–283. MR 56#345.

— [78] *Polynômes à valeurs entières et propriété de Skolem*, J. reine angew. Math., 303/304, 1978, 366–378. MR 80d:13001.

— [79a] *Anneaux de Skolem*, Archiv Math., 32, 1979, 555–568. MR 81k:13005.

— [79b] *Polynômes à valeurs entières ainsi que leurs derivées*, Ann. Sci. Univ. Clermont II, sér. math., 18, 1979, 47–64. MR 81d:13006.

— [82] *La propriéte de Skolem forte*, VI Congr. Int. Math. d'expression latine, Gauthier-Villars, 1982, 227–230. MR 83i:13019.

— [83] *Le Nullstellensatz de Hilbert et les polynômes à valeurs entières*, Monatsh. Math., 95, 1983, 181–195. MR 86j:13018.

— [87] *Un anneau de Prüfer*, J. Algebra, 107, 1987, 1–16. MR 88i:13022.

— [88] *Idéaux de polynômes et idéaux de valeurs*, Manuscr. Math., 60, 1988, 277-298. MR 89j:13007.

— [91] *Anneaux de polynômes à valeurs entières et anneaux de Prüfer*, Comptes Rendus Acad. Sci. Paris, 312, 1991, 715-720. MR 92d:13011.

— [93] *Dérivées et differences divisées à valeurs entiers*, Acta Arith., 63, 1993, 144-156.

CHABERT, J.-L., GERBOUD,G. [93] *Polynômes à valeurs entières et binômes de Fermat*, Canad. J. Math., 45, 1993, 6-26.

CHASSÉ,G. [86] *Combinatorial cycles of a polynomial map over a commutative field*, Discrete Math., 61, 1986, 21-26. MR 87k:12001.

CHATELET,F. [67] *Les idéaux de l'anneau des polynômes d'une variable à coefficients entiers*, Alg. Zahlentheorie, 43-51, Bibl. Inst. Mannheim 1967. MR 36#2611.

DICKSON,L.E. [10] *Definite forms in a finite field*, Trans. Amer. Math. Soc., 10, 1909, 109-122.

DUEBALL,F. [49] *Bestimmung von Polynomen aus ihren Werten* (mod p^n), Math. Nachr., 3, 1949, 71-76. MR 11 p.715.

EVERTSE,J.H. [84] *On equations in S-units and the Thue-Mahler equation*, Inv. math., 75, 1984, 561-584. MR 85f:11048.

FATOU,D. [19] *Sur les équations fonctionnelles*, Bull. Soc. Math. France, 47, 1919, 161-271; 48, 1920, 33-94, 208-314.

— [26] *Sur l'itération des fonctions transcendantes entières*, Acta Mathematica, 47, 1926, 337-370.

FOSTER,A.L. [67] *Semiprimal algebras: Characterization and normal-decomposition*, Math. Zeitschr., 99, 1967, 105-116. MR 35#6605.

— [70] *Congruence relations and functional completeness in universal algebras; structure theory of hemiprimals, I*, Math. Zeitschr., 113, 1970, 293-308. MR 41#5271.

FUKASAWA,S. [26] *Über ganzwertige ganze Funktionen*, Tôhoku Math. J., 27, 1926, 41-52.

— [28] *Über ganzwertige ganze Funktionen*, Tôhoku Math. J., 29, 1928,131-144.

GELFOND,A.O. [29a] *Sur un théorème de M.Pólya*, Atti della Reale Acad. Naz. dei Lincei, Rendiconti cl.sci.fis.mat.nat., (6), 10, 1929, 569-574.

— [29b] *Sur les propriétés arithmétiques des fonctions entières*, Tôhoku Math. J., 30, 1929, 280-285.

— [33] *On functions integral-valued at points of a geometric progression*, Mat. Sbornik, 40, 1933, 42-47. (Russian)

— [67] *On functions attaining integral values*, Mat. Zametki, 1, 1967, 509-513. MR 35#6630. (Russian)

GERBOUD,G. [86] *Etude de certains polynômes à valeurs entiers*, Thèse 3 cycle, Université de Provence 1986.

— [88a] *Exemples d'anneaux A pour lesquels $\left(\binom{X}{n} \right)_{n \in N}$ est une base du A-module des polynômes entières sur A*, Comptes Rendus Acad. Sci. Paris, 307, 1988, 1-4. MR 89h:13025.

116

— [88b] *Polynômes à valeurs entières sur l'anneau des entiers de Gauss*, Comptes Rendus Acad. Sci. Paris, 307, 1988, 375–378. MR 89m:11097.

— [89] *Construction sur un anneau de Dedekind d'une ,base régulière de polynômes à valeurs entières*, Manuscr. Math., 65, 1989, 167–179.   MR 90h:13016.

— [93] *Substituabilité d'un anneau de Dedekind*, Comptes Rendus Acad. Sci. Paris, 317, 1993, 29–32.

GILMER,R. [89] *Sets that determine integer-valued polynomials*, J. Number Th., 33, 1989, 95–100. MR 90g:11142.

— [90] *Prüfer domains and rings of integer-valued polynomials*, J. Algebra, 129, 1990, 502–517. MR 91b:13023.

GILMER,R., HEINZER,W., LANTZ,D. [92] *The Noetherian property in rings of integer-valued polynomials*, Trans. Amer. Math. Soc., 338, 1993, 187–199. MR 93j:13027.

GILMER,R., HEINZER,W., LANTZ,D., SMITH,W. [90] *The ring of integer-valued polynomials of a Dedekind domain*, Proc. Amer. Math. Soc., 108, 1990, 673–681. MR 90h:13017.

GILMER,R., SMITH,W.W. [83] *Finitely generated ideals of the ring of integer-valued polynomials*, J. Algebra, 81, 1983, 150–164. MR 85b:13020.

— [85] *Integer-valued polynomials and the strong two-generator property*, Houston J.Math. 11, 1985, 65–74. MR 86b:13010.

GOLDMAN,O. [51] *Hilbert rings and the Hilbert Nullstellensatz*, Math. Zeitschr., 54, 1951, 136-140. MR 13 p.427.

GRAMAIN,F. [78a] *Fonctions entières arithmétiques*, Séminaire Pierre Lelong — Henri Skoda (Analyse), Année 1976/77, Lecture Notes in Math. 694, 1978, 96–125. MR 80f:10046.

— [78b] *Fonctions entières arithmétiques*, Sém. Delange–Pisot–Poitou, 19, 1977/78, exp.8. MR 80b:10068.

— [80] *Sur le théorème de Fukasawa - Gel'fond - Gruman - Masser*, Sém. Delange–Pisot–Poitou, 1980/81, 67–86, Birkhäuser 1983. MR 85d:11071.

— [81] *Sur le théorème de Fukasawa - Gel'fond*, Inv. math., 63, 1981, 495–506. MR 83g:30028.

— [90] *Fonctions entières d'une ou plusieurs variables complexes prenant des valeurs entières sur une progression geometrique*, Cinquante ans de polynômes, Fifty years of polynomials, Lecture Notes in Math. 1415, 123–137, Springer 1990. MR 91e:11086.

GRAMAIN,F., MIGNOTTE,M. [83] *Fonctions entières arithmétiques*, Progress Math., 31, 1983, 99-124. MR 85d:30033.

GRAMAIN,F., SCHNITZER,F.J. [89] *Ganze ganzwertige Funktionen: historische Bemerkungen*, Complex methods in partial differential equations, 151–177, Akademie Verlag 1989. MR 91j:30021.

GRUMAN,L. [80] *Propriétés arithmétiques des fonctions entières*, Bull. Soc. Math. France, 108, 1980, 421–440. MR 82g:10072.

GUNJI,H., McQUILLAN D.L. [69] *On polynomials with integer coefficients*, J. Number Th., 1, 1969, 486–493, MR 40#2648.

— [70] *On a class of ideals in an algebraic number field*, J. Number Th., 2, 1970, 207–222, MR 41#1681.

— [75] *On rings with a certain divisibility property*, Michigan J. Math., 22, 1975, 289–299. MR 53#405.

— [78] *Polynomials with integral values*, Proc. Roy. Irish Acad., A78, 1978, 1–7. MR 57#12491.

HALL,R.R. [71] *On pseudo-polynomials*, Mathematika, 18, 1971, 71–77. MR 45 #3369.

HALTER-KOCH,F., NARKIEWICZ,W. [92a] *Finiteness properties of polynomial mappings*, Math. Nachr., 159, 1992, 7–18.

— [92b] *Polynomial maps with non-trivial common factor*, Sém. Th. Nombres, Bordeaux, 4, 1992, 187–198.

— [92c] *Commutative rings and binomial coefficients*, Monatsh. Math., 114, 1992, 107–110.

— [94] *Polynomial cycles in finitely generated domains*, Monatsh. Math., to appear.

HAOUAT,Y. [86] *Anneaux de polynômes à valeurs entières sur un anneau de valuation ou de Seidenberg*, Ann. Sci. Univ. Clermont-Ferrand II, sér. math., 23, 1986, 91–98. MR 88h:13010.

HAOUAT,Y.,GRAZZINI,F. [77] *Polynômes et différences divisées*, Comptes Rendus Acad. Sci. Paris, 284, 1977, A1171–A1173. MR 55#10430.

— [78] *Différences finies divisées sur un anneau S(2)*, Comptes Rendus Acad. Sci. Paris, 286, 1978, A723–A725. MR 81e:13010.

— [79] *Polynômes de Barsky*, Ann. Scient. Univ. Clermont II, sér.math., 18, 1979, 65–81. MR 81b:12019.

HARDY,G.H. [16] *On a theorem of Mr.G.Pólya*, Proc. Cambridge Phil. Soc., 19, 1916–1919, 60–63.

HEITMANN,R.C. [76] *Generating ideals in Prüfer domains*, Pacific J. Math., 62, 1976, 117–126. MR 53#10784.

HENSEL,K. [96] *Ueber den grössten gemeinsamen Theiler aller Zahlen, welche durch eine ganze Function von n Veränderlichen darstellbar sind*, J. reine angew. Math., 116, 1896, 350–356.

IVÁNYI,A. [72] *On multiplicative functions with congruence property*, Ann. Univ. Sci. Budapest, 15, 1972, 133–137. MR 48#8364.

JACOB,G. [76] *Polynômes représentant la fonction nulle sur un anneau commutatif unitaire*, Comptes Rendus Acad. Sci. Paris, 283, 1976, 421–424. MR 55#327.

— [80] *Anneau de fonctions polynômes d'un anneau commutatif unitaire*, Comm. Algebra, 1980, 793–811. MR 82j:13007.

JANKOWSKI,L., MARLEWSKI,A. [90] *On the rational polynomials having the same image of the rational number set*, Funct. et Approx., 19, 1990, 139–148. MR 92b:11012.

KELLER,G.,OLSON,F.R. [68] *Counting polynomial functions* (mod p^n), Duke Math. J., 35, 1968, 835–838. MR 38#1079.

KEMPNER,A.J. [21] *Polynomials and their residue systems*, Trans. Amer. Math. Soc., 22, 1921, 240–288.

KOCHEN,S. [69] *Integer-valued rational functions over the p-adic numbers: A p-adic analogue of the theory of real fields*, Proc. Symp. Pure Math. 12, Number Theory, AMS 1969, 57–73. MR 41#1685.

KRULL,W. [51] *Jacobsonsche Ringe, Hilbertscher Nullstellensatz, Dimensionstheorie*, Math. Zeitschr., 54, 1951, 354–387. MR 13 p.903.

KUBOTA,K.K. [72a] *Note on a conjecture of W.Narkiewicz*, J. Number Th., 4, 1972, 181–190. MR 46#9007.

— [72b] *Factors of polynomials under composition*, J. Number Th., 4, 1972, 587–595. MR 58#27922.

— [73] *Image sets of polynomials*, Acta Arith., 23, 1973, 183-194. MR 48#8459.

KUBOTA,K.K., LIARDET,P. [76] *Réfutation d'une conjecture de W.Narkiewicz*, Comptes Rendus Acad. Sci. Paris, 282, 1976, A1261-A1264. MR 54#12725.

LANDAU,E. [20] *Note on Mr. Hardy's extension of a theorem of Mr. Pólya*, Proc. Cambridge Philos. Soc., 20, 1920, 14–15 = Collected Works, VII, 257–258, Thales-Verlag.

LAOHAKASOL,V., UBOLSRI,P. [80] *A short note on integral-valued polynomials*, Southeast Asian Bull. Math., 4, 1980, 43–47, MR 83k:10036.

LENSTRA,H.W.Jr. [77] *Euclidean number fields of large degree*, Inv. math., 38, 1977, 237-254. MR 55#2836.

LEUTBECHER,A., NIKLASCH,G. [89] *On cliques of exceptional units and Lenstra's construction of Euclidean fields*, Number Theory, Lecture Notes in Math. 1380, 150–178, Springer 1989. MR 90i:11123.

LEWIS,D.J. [56] *Ideals and polynomial functions*, American J. Math, 78, 1956, 71–77. MR 19 p.526.

— [72] *Invariant sets of morphisms in projective and affine number spaces*, J. Algebra, 20, 1972, 419–434. MR 46#1746.

LEWIS,D.J., MORTON,P. [81] *Quotients of polynomials and a theorem of Pisot and Cantor*, J. Fac. Sci. Tokyo, IA, Math., 28, 1981, 813–822. MR 83f:12003.

LIARDET,P. [70] *Transformations rationnelles et ensembles algébriques*, Thése 3 cycle, Marseille 1970.

— [71] *Sur les transformations polynômiales et rationnelles*, Sém. Th. Nombres Bordeaux, 1971/72, exp. 29. MR 52#13760.

— [72] *Sur une conjecture de Narkiewicz*, Comptes Rendus Acad. Sci. Paris, 274, 1972, 1836 - 1838. MR 47#3391.

— [75] *Sur la stabilité rationnelle ou algébrique d'ensembles de nombres algébriques*, Thése, Aix-Marseille II, 1975.

LIND,D.A. [71] *Which polynomials over an algebraic number field map the algebraic integers into themselves?*, American Math. Monthly, 78, 1971, 179–180.

LINDEN,F.J. van der [88] *Integer valued polynomials over function fields*, Indag. Math., 50, 1988, 293–308. MR 90a:11142.

LOPER,A. [88] *On rings without a certain divisibility property*, J. Number Th., 28, 1988, 132–144. MR 89a:12006.

MASLEY,J.M.,MONTGOMERY,H.L. [76] *Cyclotomic fields with unique factorization*, J. reine angew. Math., 286/287, 1976, 248–256. MR 55#2834.

MASSER,D. [80] *Sur les fonctions entières*, Comptes Rendus Acad. Sci. Paris, 291, 1980, A1–A4. MR 81i:10048.

MATLIS,E. [66] *Decomposable modules*, Trans. Amer. Math. Soc., 121, 1966, 147–179. MR 34#1349.

— [70] *The two-generator property for ideals*, Michigan J. Math., 17, 1970, 257–265. MR 42#247.

McQUILLAN,D.L. [73a] *Modules over algebraic integers*, Sém. Th. Nombres, Bordeaux, 1972/73, exp.4. MR 53#5548.

— [73b] *Réseaux sur les anneaux d'entiers algébriques*, Sém. Delange–Pisot–Poitou, 14, 1972/73, n⁰.25. MR 53#13156.

— [78] *On the coefficients and values of polynomial rings*, Archiv Math., 30, 1978, 8–13. MR 57#6002.

— [85a] *On Prüfer domains of polynomials*, J. reine angew. Math., 358, 1985, 162–178. MR 86k:13019.

— [85b] *On ideals in Prüfer domains of polynomials*, Archiv Math., 45, 1985, 517–527. MR 87c:13023.

— [85c] *Rings of integer-valued polynomials determined by finite sets*, Proc. Roy. Irish Acad., 85A, 1985, 177–184. MR 87g:13016.

— [91] *On a theorem of R.Gilmer*, J. Number Th., 39, 1991, 245–250. MR 92i:13016.

MORTON.P. [92] *Arithmetic properties of periodic points of quadratic maps*, Acta Arith., 62, 1992, 343–372; II, preprint 1994.

MORTON,P., SILVERMAN,J.H. [93a] *Periodic points, multiplicities and dynamical units*, preprint 1993.

— [93b] *Rational periodic points of rational functions*, preprint 1993.

MOUSSA,P. GERONIMO,J.S., BESSIS,D. [84] *Ensembles de Julia et propriétès de localisation des familles itérées d'entiers algébriques*, Comptes Rendus Acad. Sci. Paris, 299, 1984, 281–284. MR 86f:58133.

MULLEN,G.,STEVENS,H. [84] *Polynomial functions* (mod *m*), Acta Math. Ac. Sci. Hung., 44, 1984, 237–241. MR 86a:11004.

NAGELL,T. [18] *Einige Sätze über die ganzen rationalen Funktionen*, Nyt Tydsskrift f. Matematik, 29B, 1918, 53–62.

— [19] *Über zahlentheoretische Polynome*, Norsk Mat. Tidsskrift, 1, 1919, 14–23.

NAKANO,N. [53] *Idealtheorie in einem speziellen unendlichen algebraischen Zahlkörper*, J. Sci. Hiroshima Univ., Ser.A., 16, 1953, 425–439. MR 15 p.510.

NARKIEWICZ W. [62] *On polynomial transformations*, Acta Arith., 7, 1962, 241–249. MR 26#110; II, *ibidem*,8, 1962/63, 11–19. MR 26#4987.

— [63a] *Remark on rational transformations*, Colloq. Math., 10, 1963, 139–142. MR 26#6155.

— [63b] *Problems 415, 416*, Colloq. Math., 10. 1963, p.187.

— [64] *On transformations by polynomials in two variables*, Colloq. Math., 12, 1964, 53–58. MR 29#4757; II, *ibidem*,13, 1964, 101–106. MR 30#3868.

— [65] *On polynomial transformations in several variables*, Acta Arith., 11, 1965, 163–168. MR 32#4084.

— [66] *Solution to Problem 51*, Wiadom. Mat., (2), 9, 1966, 101–102. (Polish).

— [89] *Polynomial cycles in algebraic number fields*, Colloq. Math., 58, 1989, 149–153. MR 90k:11135.

NEČAEV,A.A. [80] *Polynomial transformations of finite commutative local principal ideal rings*, Mat. Zametki, 27, 1980, 885–897. MR 82b:13013. (Russian)

NIVEN,I.,WARREN,L.J. [57] *A generalization of Fermat's theorem*, Proc. Amer. Math. Soc., 8, 1957, 306–313. MR 19 p.246.

NÖBAUER,W. [76] *Compatible and conservative functions on residue-class rings of the integers*, Coll. Math. Soc. J. Bólyai 13, Topics in Number Theory, Debrecen 1974, North-Holland 1976, 245–257. MR 55#12709.

NORTHCOTT,D.G. [49a] *An inequality in the theory of arithmetic on algebraic varieties*, Proc. Cambridge Philos. Soc., 45. 1949. 502–509. MR 11 p.390.

— [49b] *A further inequality in the theory of arithmetic on algebraic varieties*, Proc. Cambridge Philos. Soc., 45, 1949, 510–518. MR 11 p.390.

— [50] *Periodic points on an algebraic variety*, Annals of Math., 51, 1950, 167–177. MR 11 p.615.

OSTROWSKI,A. [19] *Über ganzwertige Polynome in algebraischen Zahlkörpern*, J. reine angew. Math., 149, 1919, 117–124.

PEZDA,T. [93] *Polynomial cycles in certain local domains*, Acta Arith., 66, 1994, 11–22.

— [94a] *Cycles of polynomial mappings in several variables*, Manuscr. Math., 83, 1994, 279–289.

— [94b] *Cycles of polynomials in algebraically closed fields of positive characteristics*, Colloq. Math., 67, 1994, 187–195.

PHONG,B.M. [91] *Multiplicative functions satisfying a congruence property*, Studia Sci. Math. Hungar., 26, 1991, 123–128.

PHONG,B.M., FEHÉR,J. [90] *Note on multiplicative functions satisfying a congruence property*, Ann. Univ. Sci. Budapest, 33, 1990, 261–265. MR 93d:11008.

PISOT,C. [42] *Über ganzwertige ganze Funktionen*, Jahresber. Deutsche Math. Ver., 52, 1942, 95–102. MR 4 p.270.

— [46a] *Sur les fonctions arithmétiques analytiques à croissance exponentielle*, Comptes Rendus Acad. Sci. Paris, 222, 1946, 998–990. MR 8 p.23.

— [46b] *Sur les fonctions analytiques arithmétiques et presque arithmétiques*, Comptes Rendus Acad. Sci. Paris, 222, 1946, 1027–1028. MR 8 p.23.

PÓLYA,G. [15] *Ueber ganzwertige ganze Funktionen*, Rend. Circ. Mat. Palermo, 40, 1915, 1–16.

— [19] *Über ganzwertige Polynome in algebraischen Zahlkörpern*, J. reine angew. Math., 149, 1919, 97–116.

— [20] *Ueber ganze ganzwertige Funktionen*, Nachr. Ges. Wiss. Göttingen, 1920, 1–10.

PRESTEL,A., RIPOLI,C.C. [91] *Integral-valued rational functions on valued fields*, Manuscr. Math., 73, 1991, 437–452. MR 93b:12015.

RAUSCH,U. [87] *On a class of integer-valued functions*, Archiv Math., 48, 1987, 63–67. MR 88c:11011.

RAUZY,G. [67] *Ensembles arithmétiquement dense*, Comptes Rendus Acad. Sci. Paris, 265, 1967, 37–38. MR 36#5104.

RÉDEI,L., SZELE, T. [47] *Algebraischzahlentheoretische Betrachtungen über Ringe*, I, Acta Mathematica, 79, 1947, 291–320. MR 9 p.407; II, *ibidem*,82, 1950, 209–241. MR 12 p.6.

ROBINSON,R.M. [71] *Integer-valued entire functions*, Trans. Amer. Math. Soc., 153, 1971, 451–468. MR 43#522.

ROGERS,K., STRAUS,E.G. [85] *Infinitely integer-valued polynomials over an algebraic number field*, Pacific J. Math., 118, 1985, 507–522. MR 86i:11058.

ROQUETTE,P. [71] *Bemerkungen zur Theorie der formal p-adischen Körper*, Beiträge zur Algebra und Geometrie, 1, 1971. 177–193. MR 45#8643.

ROSENBERG,I.G. [75] *Polynomial functions over finite rings*,   Glasnik Mat., 10, 1975, 25–33. MR 52#352.

RUSH,D.E. [85] *Generating ideals in rings of integer-valued polynomials*, J. Algebra, 92, 1985, 389–394. MR 86g:12007.

RUSSO,P. WALDE,R. [94] *Rational periodic points of the quadratic function* $Q_c(x) = x^2 + c$, American Math. Monthly, 101, 1994, 318–331.

RUZSA,I. [71] *On congruence-preserving functions*, Math. Lapok, 22, 1971, 125–134. MR 48#2044. (Hungarian)

SALLY,J., VASCONCELOS,W. [74] *Stable rings*, J.Pure Applied Algebra, 4, 1974, 319–336. MR 53#13185.

SATO,D., STRAUS,E.G. [64] *Rate of growth of Hurwitz entire functions and integer valued entire functions*, Bull. Amer. Math. Soc., 70, 1964, 303–307. MR 28#3161.

SCHÜLTING,H.-W. [79] *Über die Erzeugendenanzahl invertierbarer Ideale in Prüferringen*, Comm. Algebra, 7, 1979, 1331–1349. MR 81j:12016.

SELBERG,A. [41a] *Über ganzwertige ganze transzendente Funktionen*, Arch. Math. Naturvid., 44, 1941, 45–52. MR 2 p.356.

— [41b] *Über einen Satz von Gelfond*, Arch. Math.Naturvid., 44, 1941, 159–170. MR 4 p.6.

— [41c] *Über ganzwertige ganze transzendente Funktionen, II.*, Arch. Math. Naturvid., 44, 1941, 171–181. MR 4 p.6.

SHAPIRO,H.S. [57] *The range of an integer-valued polynomial*, American Math. Monthly, 64, 1957, 424–425. MR 19 p.115.

SHIBATA,F., SUGATANI,T., YOSHIDA,K. [86] *Note on rings of integral-valued polynomials*, Comptes Rendus Math. Rep. Acad. Sci. Canada, 8, 1986, 297-301. MR 87j:13024.

SILVERMAN,J.H. [93] Integer points, diophantine approximation and iteration of rational maps, Duke Math. J., 71,1993,793-829.

SINGMASTER,D. [74], *On polynomial functions mod m*, J. Number Th., 6, 1974, 345-352. MR 51#3127.

SKOLEM,T. [36] *Ein Satz über ganzwertige Polynome*, Kong. Norske Vid. Selsk., 9, 1936, 111-113.

— [37a] *Über die Lösbarkeit gewisser linearer Gleichungen im Bereiche der ganzwertigen Polynome*, Kong. Norske Vid. Selsk. Forh., 9, 1937, 134-137.

— [37b] *Sätze über ganzwertige Polynome*, Kong. Norske Vid. Selsk. Forh., 10, 1937, 12-15.

— [40] *Einige Sätze über Polynome*, Avhandl. Norske Videnskap. Akad. Oslo, I., Mat.-Naturv. Kl., 1940, No.4, 1-16. MR 2 p.247.

SOMAYAJULU,A. [68] *On arithmetic functions with congruence property*, Portugal. Math., 27, 1968, 83-85. MR 41#5281.

SPIRA,R. [68] *Polynomial interpolation over commutative rings*, American Math. Monthly, 1968, 638-640. MR 37#5199.

STÄCKEL,P. [95] *Ueber arithmetische Eigenschaften analytischer Funktionen*, Math. Ann., 46, 1895, 513-520.

STEINITZ,E. [10] *Algebraische Theorie der Körper*, J. reine angew. Math., 137, 1910, 167-309.

STRAUS,E.G. [51] *On the polynomials whose derivatives have integral values on the integers*, Proc. Amer. Math. Soc., 2, 1951, 24-27. MR 12 p.700.

— [52] *Functions periodic modulo each of a sequence of integers*, Duke Math. J., 19, 1952, 375-395. MR 14 p.21.

SUBBARAO,M.V. [66] *Arithmetic functions satisfying a congruence property*, Canad. Math. Bull., 9, 1966, 143-147. MR 33#3993.

SZEKERES,G. [52] *A canonical basis for the ideals of a polynomial domain*, American Math. Monthly, 59, 1952, 379-386. MR 13 p.903.

TROTTER,P.G. [78] *Ideals in $Z[x,y]$*, Acta Math. Ac. Sci. Hung., 32, 1978, 63-73. MR 81g:13006.

TURK,J. [86] *The fixed divisor of a polynomial*, American Math. Monthly, 93, 1986, 282-286. MR 87e:12007.

UCHIDA,K. [71] *Class numbers of imaginary abelian number fields,I.*, Tohoku Math.J. (2), 23, 1971, 97-104. MR 44#2727.

WAGNER,C.G. [76] *Polynomials over $GF(q,x)$ with integral-valued differences*, Archiv Math., 27, 1976, 495-501. MR 54# 5197.

WALDE,R.,RUSSO,P. [94] *Rational periodic points of the quadratic function $Q_c(x) = x^2 + c$*, American Math. Monthly, 101, 1994, 318-331.

WALDSCHMIDT,M. [78] *Pólya's theorem by Schneider's method*. Acta Math. Ac. Sci. Hung., 31, 1978, 21-25. MR 58#5542.

WALLISER,R. [69] *Verallgemeinerte ganze ganzwertige Funktionen vom Exponentialtypus*, J. reine angew. Math., 235, 1969, 189-206. MR 39# 5797.

— [85] *Über ganze Funktionen, die in einer geometrischen Folge ganze Werte annehmen*, Monatsh. Math., 100, 1985, 329–335. MR 87b:30043.

WIESENBAUER,J. [82] *On polynomial functions over residue class rings of Z*, Contributions to General Algebra 2, 1982, 395–398. MR 85b:11114.

ZANTEMA,H. [82] *Integer-valued polynomials over a number field*, Manuscr. Math., 40, 1982, 155–203. MR 84i:12005.

INDEX

Springer-Verlag
and the Environment

We at Springer-Verlag firmly believe that an international science publisher has a special obligation to the environment, and our corporate policies consistently reflect this conviction.

We also expect our business partners – paper mills, printers, packaging manufacturers, etc. – to commit themselves to using environmentally friendly materials and production processes.

The paper in this book is made from low- or no-chlorine pulp and is acid free, in conformance with international standards for paper permanency.

Printing: Weihert-Druck GmbH, Darmstadt
Binding: Theo Gansert Buchbinderei GmbH, Weinheim

Lecture Notes in Mathematics

For information about Vols. 1–1419
please contact your bookseller or Springer-Verlag

Vol. 1460: G. Toscani, V. Boffi, S. Rionero (Eds.), Mathematical Aspects of Fluid Plasma Dynamics. Proceedings, 1988. V, 221 pages. 1991.

Vol. 1461: R. Gorenflo, S. Vessella, Abel Integral Equations. VII, 215 pages. 1991.

Vol. 1462: D. Mond, J. Montaldi (Eds.), Singularity Theory and its Applications. Warwick 1989, Part I. VIII, 405 pages. 1991.

Vol. 1463: R. Roberts, I. Stewart (Eds.), Singularity Theory and its Applications. Warwick 1989, Part II. VIII, 322 pages. 1991.

Vol. 1464: D. L. Burkholder, E. Pardoux, A. Sznitman, Ecole d'Eté de Probabilités de Saint- Flour XIX-1989. Editor: P. L. Hennequin. VI, 256 pages. 1991.

Vol. 1465: G. David, Wavelets and Singular Integrals on Curves and Surfaces. X, 107 pages. 1991.

Vol. 1466: W. Banaszczyk, Additive Subgroups of Topological Vector Spaces. VII, 178 pages. 1991.

Vol. 1467: W. M. Schmidt, Diophantine Approximations and Diophantine Equations. VIII, 217 pages. 1991.

Vol. 1468: J. Noguchi, T. Ohsawa (Eds.), Prospects in Complex Geometry. Proceedings, 1989. VII, 421 pages. 1991.

Vol. 1469: J. Lindenstrauss, V. D. Milman (Eds.), Geometric Aspects of Functional Analysis. Seminar 1989-90. XI, 191 pages. 1991.

Vol. 1470: E. Odell, H. Rosenthal (Eds.), Functional Analysis. Proceedings, 1987-89. VII, 199 pages. 1991.

Vol. 1471: A. A. Panchishkin, Non-Archimedean L-Functions of Siegel and Hilbert Modular Forms. VII, 157 pages. 1991.

Vol. 1472: T. T. Nielsen, Bose Algebras: The Complex and Real Wave Representations. V, 132 pages. 1991.

Vol. 1473: Y. Hino, S. Murakami, T. Naito, Functional Differential Equations with Infinite Delay. X, 317 pages. 1991.

Vol. 1474: S. Jackowski, B. Oliver, K. Pawałowski (Eds.), Algebraic Topology, Poznań 1989. Proceedings. VIII, 397 pages. 1991.

Vol. 1475: S. Busenberg, M. Martelli (Eds.), Delay Differential Equations and Dynamical Systems. Proceedings, 1990. VIII, 249 pages. 1991.

Vol. 1476: M. Bekkali, Topics in Set Theory. VII, 120 pages. 1991.

Vol. 1477: R. Jajte, Strong Limit Theorems in Noncommutative L_2-Spaces. X, 113 pages. 1991.

Vol. 1478: M.-P. Malliavin (Ed.), Topics in Invariant Theory. Seminar 1989-1990. VI, 272 pages. 1991.

Vol. 1479: S. Bloch, I. Dolgachev, W. Fulton (Eds.), Algebraic Geometry. Proceedings, 1989. VII, 300 pages. 1991.

Vol. 1480: F. Dumortier, R. Roussarie, J. Sotomayor, H. Żoładek, Bifurcations of Planar Vector Fields: Nilpotent Singularities and Abelian Integrals. VIII, 226 pages. 1991.

Vol. 1481: D. Ferus, U. Pinkall, U. Simon, B. Wegner (Eds.), Global Differential Geometry and Global Analysis. Proceedings, 1991. VIII, 283 pages. 1991.

Vol. 1482: J. Chabrowski, The Dirichlet Problem with L^2-Boundary Data for Elliptic Linear Equations. VI, 173 pages. 1991.

Vol. 1483: E. Reithmeier, Periodic Solutions of Nonlinear Dynamical Systems. VI, 171 pages. 1991.

Vol. 1484: H. Delfs, Homology of Locally Semialgebraic Spaces. IX, 136 pages. 1991.

Vol. 1485: J. Azéma, P. A. Meyer, M. Yor (Eds.), Séminaire de Probabilités XXV. VIII, 440 pages. 1991.

Vol. 1486: L. Arnold, H. Crauel, J.-P. Eckmann (Eds.), Lyapunov Exponents. Proceedings, 1990. VIII, 365 pages. 1991.

Vol. 1487: E. Freitag, Singular Modular Forms and Theta Relations. VI, 172 pages. 1991.

Vol. 1488: A. Carboni, M. C. Pedicchio, G. Rosolini (Eds.), Category Theory. Proceedings, 1990. VII, 494 pages. 1991.

Vol. 1489: A. Mielke, Hamiltonian and Lagrangian Flows on Center Manifolds. X, 140 pages. 1991.

Vol. 1490: K. Metsch, Linear Spaces with Few Lines. XIII, 196 pages. 1991.

Vol. 1491: E. Lluis-Puebla, J.-L. Loday, H. Gillet, C. Soulé, V. Snaith, Higher Algebraic K-Theory: an overview. IX, 164 pages. 1992.

Vol. 1492: K. R. Wicks, Fractals and Hyperspaces. VIII, 168 pages. 1991.

Vol. 1493: E. Benoît (Ed.), Dynamic Bifurcations. Proceedings, Luminy 1990. VII, 219 pages. 1991.

Vol. 1494: M.-T. Cheng, X.-W. Zhou, D.-G. Deng (Eds.), Harmonic Analysis. Proceedings, 1988. IX, 226 pages. 1991.

Vol. 1495: J. M. Bony, G. Grubb, L. Hörmander, H. Komatsu, J. Sjöstrand, Microlocal Analysis and Applications. Montecatini Terme, 1989. Editors: L. Cattabriga, L. Rodino. VII, 349 pages. 1991.

Vol. 1496: C. Foias, B. Francis, J. W. Helton, H. Kwakernaak, J. B. Pearson, H_∞-Control Theory. Como, 1990. Editors: E. Mosca, L. Pandolfi. VII, 336 pages. 1991.

Vol. 1497: G. T. Herman, A. K. Louis, F. Natterer (Eds.), Mathematical Methods in Tomography. Proceedings 1990. X, 268 pages. 1991.

Vol. 1498: R. Lang, Spectral Theory of Random Schrödinger Operators. X, 125 pages. 1991.

Vol. 1499: K. Taira, Boundary Value Problems and Markov Processes. IX, 132 pages. 1991.

Vol. 1500: J.-P. Serre, Lie Algebras and Lie Groups. VII, 168 pages. 1992.

Vol. 1501: A. De Masi, E. Presutti, Mathematical Methods for Hydrodynamic Limits. IX, 196 pages. 1991.

Vol. 1502: C. Simpson, Asymptotic Behavior of Monodromy. V, 139 pages. 1991.

Vol. 1503: S. Shokranian, The Selberg-Arthur Trace Formula (Lectures by J. Arthur). VII, 97 pages. 1991.

Vol. 1504: J. Cheeger, M. Gromov, C. Okonek, P. Pansu, Geometric Topology: Recent Developments. Editors: P. de Bartolomeis, F. Tricerri. VII, 197 pages. 1991.

Vol. 1505: K. Kajitani, T. Nishitani, The Hyperbolic Cauchy Problem. VII, 168 pages. 1991.

Vol. 1506: A. Buium, Differential Algebraic Groups of Finite Dimension. XV, 145 pages. 1992.

Vol. 1507: K. Hulek, T. Peternell, M. Schneider, F.-O. Schreyer (Eds.), Complex Algebraic Varieties. Proceedings, 1990. VII, 179 pages. 1992.

Vol. 1508: M. Vuorinen (Ed.), Quasiconformal Space Mappings. A Collection of Surveys 1960-1990. IX, 148 pages. 1992.